Research Opportunities for
Deactivating
and
Decommissioning
Department of Energy Facilities

Committee on Long-Term Research Needs for Deactivation and Decommissioning at Department of Energy Sites

Board on Radioactive Waste Management

Division on Earth and Life Studies

National Research Council

NATIONAL ACADEMY PRESS
Washington, D.C.

NOTICE: The project that is the subject of this report was approved by the Governing Board of the National Research Council, whose members are drawn from the councils of the National Academy of Sciences, the National Academy of Engineering, and the Institute of Medicine. The members of the committee responsible for the report were chosen for their special competences and with regard for appropriate balance.

Support for this study was provided by the U.S. Department of Energy under Grant No. DE-FC01-99EW59049. All opinions, findings, conclusions, or recommendations expressed herein are those of the authors and do not necessarily reflect the views of the U.S. Department of Energy.

International Standard Book Number 0309-07595-5

Additional copies of this report are available from National Academy Press, 2101 Constitution Avenue, N.W., Lockbox 285, Washington, D.C. 20055; (800) 624-6242 or (202) 334-3313 (in the Washington metropolitan area); Internet: http://www.nap.edu

Cover: Presently, deactivation and decommissioning work requires hands-on labor in hazardous areas. Science can provide new technologies that enhance worker safety and reduce cleanup time and cost. Photograph courtesy of the U.S. Department of Energy.

Copyright 2001 by the National Academy of Sciences. All rights reserved.

Printed in the United States of America

THE NATIONAL ACADEMIES

National Academy of Sciences
National Academy of Engineering
Institute of Medicine
National Research Council

The **National Academy of Sciences** is a private, nonprofit, self-perpetuating society of distinguished scholars engaged in scientific and engineering research, dedicated to the furtherance of science and technology and to their use for the general welfare. Upon the authority of the charter granted to it by the Congress in 1863, the Academy has a mandate that requires it to advise the federal government on scientific and technical matters. Dr. Bruce M. Alberts is president of the National Academy of Sciences.

The **National Academy of Engineering** was established in 1964, under the charter of the National Academy of Sciences, as a parallel organization of outstanding engineers. It is autonomous in its administration and in the selection of its members, sharing with the National Academy of Sciences the responsibility for advising the federal government. The National Academy of Engineering also sponsors engineering programs aimed at meeting national needs, encourages education and research, and recognizes the superior achievements of engineers. Dr. Wm. A. Wulf is president of the National Academy of Engineering.

The **Institute of Medicine** was established in 1970 by the National Academy of Sciences to secure the services of eminent members of appropriate professions in the examination of policy matters pertaining to the health of the public. The Institute acts under the responsibility given to the National Academy of Sciences by its congressional charter to be an adviser to the federal government and, upon its own initiative, to identify issues of medical care, research, and education. Dr. Kenneth I. Shine is president of the Institute of Medicine.

The **National Research Council** was organized by the National Academy of Sciences in 1916 to associate the broad community of science and technology with the Academy's purposes of furthering knowledge and advising the federal government. Functioning in accordance with general policies determined by the Academy, the Council has become the principal operating agency of both the National Academy of Sciences and the National Academy of Engineering in providing services to the government, the public, and the scientific and engineering communities. The Council is administered jointly by both Academies and the Institute of Medicine. Dr. Bruce M. Alberts and Dr. Wm. A. Wulf are chairman and vice chairman, respectively, of the National Research Council.

COMMITTEE ON LONG-TERM RESEARCH NEEDS FOR DEACTIVATION AND DECOMMISSIONING AT DEPARTMENT OF ENERGY SITES

PHILIP CLARK, SR., Chair, GPU Nuclear Corporation (retired), Boonton, New Jersey
ANTHONY J. CAMPILLO, Naval Research Laboratory, Washington, D.C.
FRANK CRIMI, Lockheed Martin Advanced Environmental Systems Company (retired), Saratoga, California
KEN CZERWINSKI, Massachusetts Institute of Technology, Cambridge
RACHEL DETWILER, Construction Technology Laboratories, Inc., Skokie, Illinois
HARRY HARMON, Battelle, PNNL, Savannah River Site, Aiken, South Carolina
VINCENT MASSAUT, CEN.SCK, Mol, Belgium
ALAN PENSE, Lehigh University, Bethlehem, Pennsylvania
GARY SAYLER, University of Tennessee, Knoxville
DELBERT TESAR, University of Texas, Austin

STAFF

JOHN R. WILEY, Senior Staff Officer
SUSAN B. MOCKLER, Research Associate
TONI GREENLEAF, Administrative Associate
LATRICIA C. BAILEY, Senior Project Assistant

BOARD ON RADIOACTIVE WASTE MANAGEMENT

JOHN F. AHEARNE, Chair, Sigma Xi and Duke University, Research Triangle Park, North Carolina
CHARLES MCCOMBIE, Vice-Chair, Consultant, Gipf-Oberfrick, Switzerland
ROBERT M. BERNERO, U.S. Nuclear Regulatory Commission (retired), Gaithersburg, Maryland
ROBERT J. BUDNITZ, Future Resources Associates, Inc., Berkeley, California
GREGORY R. CHOPPIN, Florida State University, Tallahassee
RODNEY EWING, University of Michigan, Ann Arbor
JAMES H. JOHNSON, JR., Howard University, Washington, D.C.
ROGER E. KASPERSON, Stockholm Environment Institute, Stockholm, Sweden
NIKOLAY LAVEROV, Russian Academy of Sciences, Moscow
JANE C. S. LONG, Mackay School of Mines, University of Nevada, Reno
ALEXANDER MACLACHLAN, E.I. du Pont de Nemours & Company (retired), Wilmington, Delaware
WILLIAM A. MILLS, Oak Ridge Associated Universities (retired), Olney, Maryland
MARTIN J. STEINDLER, Argonne National Laboratory (retired), Downers Grove, Illinois
ATSUYUKI SUZUKI, University of Tokyo, Japan
JOHN J. TAYLOR, Electric Power Research Institute (retired), Palo Alto, California
VICTORIA J. TSCHINKEL, Landers and Parsons, Tallahassee, Florida

STAFF

KEVIN D. CROWLEY, Director
MICAH D. LOWENTHAL, Staff Officer
BARBARA PASTINA, Staff Officer
GREGORY H. SYMMES, Senior Staff Officer
JOHN R. WILEY, Senior Staff Officer
SUSAN B. MOCKLER, Research Associate
DARLA J. THOMPSON, Senior Project Assistant/Research Assistant
TONI GREENLEAF, Administrative Associate
LATRICIA C. BAILEY, Senior Project Assistant
LAURA D. LLANOS, Senior Project Assistant
ANGELA R. TAYLOR, Senior Project Assistant
JAMES YATES, JR., Office Assistant

Preface

At the beginning of the nuclear era over 60 years ago, the United States was gravely concerned about its national security. The U.S. Army Corps of Engineers and later the Atomic Energy Commission (AEC) began a massive construction program to build facilities that would manufacture nuclear materials for the nation's defense. Emphasis on nuclear materials production continued throughout the Cold War. Altogether the AEC and its successor, the Department of Energy (DOE), built and maintained some 20,000 facilities (buildings that housed equipment) at sites throughout the country.

When the Cold War abruptly ended, DOE halted most nuclear materials production. Major facilities were closed, often with large inventories of radioactive materials still in process. Today many former production facilities have been declared surplus and turned over to the DOE Office of Environmental Management (EM) to be deactivated and decommissioned as part of EM's overall site cleanup activities. Deactivation includes putting the facility in a safe, stable, and monitored condition; decommissioning includes decontaminating the facility and permanently retiring the building from DOE—perhaps demolishing it or reusing it for another purpose.

In 1995, Congress chartered DOE's Environmental Management Science Program (EMSP) to bring the nation's scientific infrastructure to bear on EM's most difficult, long-term cleanup challenges. The EMSP provides grants to investigators in industry, national laboratories, and universities to undertake research that may help address these cleanup challenges. On several occasions the EMSP has asked the National Academies for advice on developing its research agenda. This report resulted from a 15-month study by an Academies committee on long-term research needs for deactivation and decommissioning (D&D) at DOE sites.

In conducting this study, the committee held six meetings and visited three DOE sites to witness ongoing D&D work, view facilities that will pose the greatest challenges for future D&D, and receive presentations from DOE and site contractor personnel. A great deal of effort was required to arrange these visits and presentations for the committee's benefit, and we recognize Mark Gilbertson and Ker-Chi Chang of DOE headquarters for their help. The visits and presentations at the DOE sites were very informative and well prepared. Our visit coordinators at the sites were John Sands, DOE, and Kim Koegler, Bechtel Hanford, at Hanford, Washington; Daryl Green, DOE, at Oak Ridge, Tennessee; and Karen Lutz, DOE, at Rocky Flats, Colorado. All of the presenters listed in Appendix A provided frank and insightful information during our meetings and site visits.

We also recognize the staff of the National Academies' Board on Radioactive Waste Management (BRWM) for their assistance during the study. John Wiley, who served as study director, helped to guide the committee through its fact finding, report writing, and report review. Kevin Crowley, BRWM director, and John Taylor, BRWM liaison, provided much helpful advice. Staff members Latricia Bailey and Toni Greenleaf were always efficient and cheerful as they handled all of the many logistic details for the committee.

Finally, I want to thank the members of the committee. They were a pleasure to work with, and each made significant contributions. The fact that a majority are active researchers was particularly helpful.

Philip Clark, Sr.
Chair

List of Report Reviewers

This report has been reviewed in draft form by individuals chosen for their diverse perspectives and technical expertise, in accordance with procedures approved by the National Research Council (NRC) Report Review Committee. The purpose of this independent review is to provide candid and critical comments that will assist the institution in making the published report as sound as possible and to ensure that the report meets institutional standards for objectivity, evidence, and responsiveness to the study charge. The content of the review comments and draft manuscript remains confidential to protect the integrity of the deliberative process. We wish to thank the following individuals for their participation in the review of this report:

Gregory Choppin, Florida State University
John Coats, Southern Illinois University
Michael Corradini, University of Wisconsin
John Evans, National Institute of Standards and Technology
Edward Lazo, OECD Nuclear Energy Agency
Alexander MacLachlan, DuPont (retired)
Raymond Wymer, Oak Ridge National Laboratory (retired)

Although the reviewers listed above have provided many constructive comments and suggestions, they were not asked to endorse the conclusions or recommendations, nor did they see the final draft of the report before its release. The review of this report was overseen by Richard Conway, appointed by the National Research Council, who was responsible for making certain that an independent examination of this report was carried out in accordance with NRC procedures and that all review comments were carefully considered. Responsibility for the final content of this report rests entirely with the authoring committee and the NRC.

Contents

EXECUTIVE SUMMARY		**1**
1	**INTRODUCTION, BACKGROUND, AND TASK**	**9**
2	**D&D CHALLENGES IN THE DOE COMPLEX**	**13**
3	**CURRENT D&D SCIENCE AND TECHNOLOGY PROGRAMS**	**33**
4	**RESEARCH RECOMMENDATIONS**	**49**
5	**PROGRAMMATIC RECOMMENDATIONS**	**71**
REFERENCES		**81**

APPENDIXES

A	Presentations to the Committee	93
B	Biographical Sketches of Committee Members	97
C	Interim Report	103
D	Illustrative Science Base and Scope for Remote Technology for Decontamination and Decommissioning of DOE Nuclear Facilities	121
E	Acronyms	135

Executive Summary

The National Academies' National Research Council (NRC) undertook this study in response to a request from the Assistant Secretary for Environmental Management (EM) of the Department of Energy (DOE) for advice on research that could lead to new approaches for deactivation and decommissioning (D&D) of surplus nuclear facilities at DOE sites.[1] DOE constructed over 20,000 such facilities throughout the country to support nuclear weapons production and other activities. Most of these facilities will eventually undergo D&D as part of EM's site cleanup mission. EM has estimated that use of new technologies may save about half of the $30 billion that it currently estimates as the cost for facility D&D. New technologies are also necessary to reduce hazards to workers, especially in D&D of the more highly contaminated and complex facilities.

The Environmental Management Science Program (EMSP) was chartered by Congress to bring the nation's science infrastructure to bear on EM's cleanup mission by funding basic, mission-oriented research. The objective of this report is to provide recommendations to the EMSP on the development of a long-term research agenda that may lead to new technologies for D&D of the complex, highly contaminated facilities formerly used by DOE. The study committee was asked to

- identify significant D&D problems that cannot be addressed effectively with current technologies and
- recommend areas of research where the EMSP can make significant contributions to solving these problems and add to scientific knowledge generally.

---

[1] In the context of this report, facilities are buildings and the equipment inside the buildings. Deactivation involves placing a facility in a safe shutdown condition, and decommissioning includes actions such as decontamination or dismantling at the end of the life of a facility to retire it from DOE service. See Chapter 2.

In recommending specific areas of research the committee was asked to take into account, where possible, the agendas of other D&D-related research programs. The statement of task also suggested that the committee make recommendations, as appropriate, on the processes by which future research needs can be identified and successful research results can be applied to DOE's D&D problems.

In addition, in his two presentations to the committee, Mark Gilbertson, Director of EM's Office of Basic and Applied Research, asked the committee to suggest a vision for the EMSP.

Facilities That Pose Significant D&D Problems

In its fact finding the committee visited DOE's Hanford, Oak Ridge, and Rocky Flats sites. It also received presentations from the Savannah River Site and the Idaho National Engineering and Environmental Laboratory. The committee found that the DOE facilities that are likely to pose the greatest challenges for future D&D are those formerly used for

- radiochemical processing of irradiated nuclear fuel and target assemblies (for production of plutonium, tritium, and other nuclear materials);
- uranium enrichment by gaseous diffusion;
- plutonium processing; and
- tritium processing.

The committee found that, while current D&D technologies probably can be made to work in the D&D of these facilities, there are opportunities to do the job more safely and effectively by developing and using new technologies. Many current technologies are labor intensive and time consuming. Most current D&D technologies require hands-on contact by workers who must operate powerful equipment (e.g., torches, saws, and lifting devices) while wearing bulky protective clothing. The facilities listed above will present hazards to workers that include penetrating radiation, airborne contamination, toxic chemicals, and other industrial hazards (see Chapter 2). Within some processing cells the levels of radiation are potentially lethal. Most of these facilities are also problematic in terms of their number, size, and mass. For example, a typical radiochemical processing facility is 1000 feet long with four-foot thick concrete shielding walls. The committee concluded that:

There are strong safety and economic incentives for developing and using innovative D&D technologies that may be achieved through scientific research. The long time frame for completing D&D (50 years or more) allows for substantive research to be completed and applied.

Research Recommendations

The EMSP should focus on long-range basic research targeted on broad (site wide) or major (essential to one or a few sites) D&D needs.

Research projects should address significant long-term problems to advance the state of knowledge well beyond the next decade. This approach maintains the EMSP long-term mission. Nevertheless, opportunities for research that provides high potential payoff in addressing urgent near-term needs may arise. As a practical matter, the EMSP may well encounter a range of research opportunities that span short- and long-term needs as well as provide for contingent approaches for D&D.

The committee found that there are four major areas of research where the EMSP can make significant contributions to solving D&D problems and contributing to scientific knowledge. These areas are

- characterization of contaminated materials,
- decontamination of equipment and facilities,
- remote intelligent systems to enhance worker safety, and
- end-state definition for facility D&D.

Characterization

Characterization of contaminated materials is necessary at several stages of D&D. Substantial cost savings may result from basic research toward developing the means, preferably real-time, minimally invasive, and field usable, to locate, identify, and quantify contaminants difficult to measure in construction materials such as concrete and stainless steel, on equipment, and in packaged wastes. The committee made three recommendations in the area of characterization.

• *Basic research leading to ultra-sensitive devices for rapid characterization and certification of amounts of radionuclides and EPA-listed substances on the surfaces of construction materials and equipment (e.g., pumps, motors).*

• *Basic research leading to development of real-time and minimally invasive methods to characterize radionuclides and EPA-listed*

substances as a function of depth in construction materials, especially concrete.
- *Basic research leading to the development of methods for remotely mapping radionuclides and EPA-listed substances.*

Decontamination

Decontamination of equipment and facilities is necessary at several stages of the D&D process to minimize the radiation exposure to personnel and to reduce the volume of radioactive waste generated. Scientific understanding of the interactions among contaminants and construction materials is required to develop more effective decontamination technologies. The committee made two recommendations in the area of decontamination.

- *Basic research toward fundamental understanding of the chemical and physical interactions of important contaminants with the primary materials of interest in D&D projects, including concrete, stainless steel, paints, and strippable coatings. Results should be used to develop first-principle models that describe the interactions and can be used to investigate improved approaches to decontamination.*
- *Basic research on biotechnological means to remove contaminants from surfaces and from within porous materials found in surplus DOE facilities.*

Robotics

Industrial safety is a major issue in D&D projects because many current technologies require hands-on labor in hazardous areas. The major opportunities for reducing risks to workers lie in development of intelligent remote systems (robots) that can substitute for human workers in hazardous areas.

The committee recommends basic research toward creating intelligent remote systems that can adapt to a variety of tasks and be readily assembled from standardized modules, with special emphasis on actuators, universal operational software, and virtual presence.

End States

The definition of end states for D&D facilities and of standards for release of materials for recycling will have a major impact on cost, schedule, and risks to the public, workers, and the environment.

However, there is insufficient scientific basis for comparing the safety of various end states.

Research should be directed toward understanding the fate and behavior of treated and untreated contaminated material by determining the fundamental chemical species of the contaminants in the host material and how the species behave. The effect of time and changing ambient conditions should be considered in these investigations. Further research should be directed at incorporating these results into risk assessments to evaluate and compare the long-term safety provided by different end-state options.

Programmatic Recommendations

The statement of task asked the committee to take into account, where possible, the agendas of other D&D-related research programs in recommending areas for research. It also suggested that the committee consider and make recommendations on identifying science needs and applying successful research results to DOE's D&D problems.

Although many areas of basic research are potentially relevant to D&D, the committee did not find any U.S. agencies that fund targeted D&D research in a way that is comparable to the EMSP. There are opportunities for the EMSP to cooperate in international D&D research programs (see Chapter 3).

Through site visits and presentations the committee developed its own judgments on research needs and opportunities (see Chapter 4). However, the best that this committee or any other advisory committee can do is to provide a snapshot overview. In-depth understanding of research needs and the maturing of promising research results to deployable technologies can be achieved only by those who live with D&D challenges on a sustained basis. The EMSP's best opportunities for in-depth identification of science needs and development of promising research results lie with site contractors and involved scientists.

International Cooperation

There are opportunities for greater cooperation between the EMSP and foreign research programs in the area of D&D. Cooperation or partnerships could save money and time and could even increase the topics that can be addressed. Cooperation may also open doors for marketing U.S. technologies abroad.

The EMSP should pursue partnerships or cooperation in international research programs. These interactions should include information sharing, conferences, jointly funded research projects, and exchange of personnel at the scientific staff level.

Science Needs Identification

There is no complete, comprehensive, and coordinated definition of D&D science needs for the DOE complex. The committee saw a variety of needs lists, particularly those from the site technology coordinating groups. Most lists were too narrow, short term, or site specific to help determine where basic research could be helpful. Presentations made to the committee indicated that most attention and funding are aimed at short-term, site-specific D&D problems. While relevance to site-specific problems is important, too narrow a focus may preclude funding novel outside-the-box research with potentially greater impact on D&D.

- *EM should encourage contractors and DOE site management to take a broader, long-term perspective of D&D needs for work to be performed in 10 years or more, so that technology solutions can be developed that provide greatly improved D&D operational capabilities.*
- *The EMSP should engage the scientific community more effectively in identifying and participating in promising areas of research. A broad sustained effort by the scientific community to understand and engage in the D&D challenges would allow them to define the scientific information needed and propose relevant topics for research. This would also attract young scientists and engineers to the D&D field at a time when the availability of scientific personnel is declining.*

Closer interaction between contractors and scientists, for example at the EMSP's biannual workshops, would be beneficial. One of the outcomes of a dialogue between contractors and researchers might be recognition of new and innovative approaches to research issues that would otherwise not be identified.

Application of Successful Research Results

The EM Office of Science and Technology (OST) faces significant obstacles in promoting the deployment of new technologies, especially in the D&D area. Most presentations to the committee during its site visits expressed the view that current technologies are adequate for the D&D of DOE's facilities—that there are no substantial technology gaps.

The committee found that most of these technologies are labor intensive, time consuming, and therefore expensive. The hands-on nature of current technologies risks exposing workers to radiation, hazardous materials, and industrial hazards.

The OST should increase efforts to transition basic research to a deployable product by improving communications and cooperation among the researchers, DOE laboratories, and contractors performing D&D. Incentives for deployment of newly developed technologies that promise advantages over older technologies should be included in D&D contracts.

EMSP Vision

According to the EMSP's congressional charter (see Chapter 3), the committee believes that the vision must extend to the applications of science and technology that create major benefits for the EM program. The committee's proposed vision statement for the EMSP is as follows:

Provide scientific knowledge to allow dramatic improvements in worker safety, cost, and schedule for meeting the national need to clean up DOE sites while protecting public health and the environment. In doing this, the EMSP will be recognized as a key partner by the focus areas and DOE sites, will be supported by Congress and stakeholders, and will be preparing and developing qualified scientists for future DOE program needs.

To help achieve this vision the committee believes that the EMSP and the rest of the EM community should develop and pursue aggressive, shared goals for improvements in worker safety, cost, and schedule. The ambitious goals for dramatic improvements (factors of three to ten) set forth in the Robotics and Intelligent Machines Roadmap (Sandia, 1998) are appropriate examples for the EMSP.[2]

The committee believes that establishing and pursuing such shared goals will significantly enhance the EMSP's efforts to identify needs and apply results. The EMSP can strengthen itself by emphasizing its role in developing new scientists and engineers and the commercial value of the research that it funds.

---

[2]These goals are not the result of the committee's analysis, nor intended to be required of contractors, but are indicative of goals that could be achievable and will help identify true breakthrough areas for research.

1
Introduction, Background, and Task

The Department of Energy's (DOE's) Environmental Management Science Program (EMSP) was created by the 104th Congress[1] to bring the nation's basic science infrastructure to bear on the massive environmental cleanup effort now under way in the DOE complex. According to DOE, the mission of the EMSP is to develop and fund a targeted, long-term research program that will result in transformational or breakthrough approaches for solving the department's environmental problems. The goal is to support research that will:

- Lead to significantly lower cleanup costs and reduced risks to workers, the public, and the environment over the long term.
- Bridge the gap between broad fundamental research that has wide-ranging applicability . . . and needs-driven applied technology.
- Serve as a stimulus for focusing the nation's science infrastructure on critical national environmental management problems. (DOE, 2000g, pp. 1-2).

To meet these objectives the EMSP provides three-year awards to investigators in industry, national laboratories, and universities to undertake research on problems relevant to DOE cleanup efforts. Project awards are competitive and are made on the basis of merit and relevance reviews managed through a partnership between the DOE Office of Environmental Management (EM), which has the primary responsibility for the cleanup mission, and the DOE Office of Science,[2] which manages DOE basic research programs. Since its establishment by Congress, the program has held five proposal competitions and has

---

[1] Public Law 104-46, 1995.
[2] Formerly the Office of Energy Research.

awarded about $265 million in funding, which puts it among the largest environmental research efforts in the federal government.

Shortly after the program was established, DOE requested advice on its structure and management from the National Academies. In response, the National Academies established the Committee on Building an Effective Environmental Management Science Program, which operated from May 1996 through March 1997. One of the primary recommendations made by this committee was that DOE should

> *develop a science plan for the EMSP. This science plan should provide a comprehensive list of significant cleanup problems in the nation's nuclear weapons complex that can be addressed through basic research and a strategy for addressing them (NRC, 1997, p. 3).*

That committee also recommended a near-term and a long-term process for developing this science plan: For the near term, program managers should develop a science plan from DOE documents. For the longer term, DOE should consult with its problem holders (i.e., site technical staff, managers, and stakeholder advisory groups that have knowledge of cleanup issues) about cleanup problems that cannot be resolved practically or efficiently with current knowledge or technologies.

As one means of implementing the longer-term recommendation, Gerald Boyd, then Acting Deputy Assistant Secretary of the Office of Science and Technology (OST), requested that the National Academies convene a committee of experts in the spring of 1998. This committee advised DOE on its first science plan for the EMSP, which DOE had decided would address subsurface contamination. Following that initial study, Carolyn Huntoon, Assistant Secretary for Environmental Management, requested two additional studies. These studies would advise DOE on developing research agendas in two areas: high-level radioactive waste and deactivating and decommissioning nuclear facilities. In response, two committees were formed under the auspices of the Board on Radioactive Waste Management. This report provides the advice requested by DOE on facility deactivation and decommissioning (D&D).

Statement of Task

The statement of task for this study (see Sidebar 1.1) charged the committee to provide recommendations for a science research program for D&D problems at DOE sites. The committee was asked to identify areas of research where the program could make significant contributions to future D&D efforts and add to scientific knowledge generally.

SIDEBAR 1.1 STATEMENT OF TASK

The objective of this study is to provide recommendations to DOE's EM Science Program on the development of a long-term basic research agenda that may lead to new technologies for the deactivation and decommissioning (D&D) of complex, highly contaminated facilities formerly used for the production of nuclear materials. The report will accomplish the following:

- Identify significant D&D problems that cannot be addressed effectively with current technologies.
- Recommend areas of research where the EM Science Program can make significant contributions to solving these problems and adding to scientific knowledge generally.

In recommending specific areas of research, the committee should take into account, where possible, the agendas of other D&D-related research programs.

The committee may also consider and make recommendations, as appropriate, on the processes by which (1) future research needs can be identified, and (2) successful research results can be applied to DOE's D&D problems.

The committee was also invited to provide recommendations on processes for identifying future D&D research needs and for applying the results of successful research to DOE's D&D problems.

On beginning this study, the committee was aware of the findings of previous studies directed at EM's cleanup task and the EMSP. The NRC Committee on Building an Environmental Management Science Program stated that "Many of EM's cleanup problems cannot be solved or even managed efficiently and safely with current technologies, in part owing to their tremendous size and scope. . . . Simply put, new technologies are required to deal with EM's most difficult problems, and new technologies demand new science" (NRC, 1997, pp. 1-2). In addition, a previous study of D&D technology development programs within OST found that, while incremental improvements of commercially available technologies are useful for near-term D&D problems, identification and development of truly new, innovative technologies needed for long-term problems were not being done effectively and new approaches were needed (NRC, 1998a).

The committee held six meetings between March 2000 and January 2001 to gather information on the most significant long-term D&D challenges at five major DOE sites and to develop this report.[3] This fact

---

[3]See Appendix A for a summary of the committee's fact-finding activities.

finding included briefings on D&D plans and challenges at the Hanford Site (Washington), Idaho National Engineering and Environmental Laboratory, Oak Ridge Site (Tennessee), Savannah River Site (South Carolina), and Rocky Flats Environmental Technology Site (Colorado). The committee toured the Hanford, Oak Ridge, and Rocky Flats sites to observe the facilities and obtain briefings from site personnel.

The committee focused primarily on scientific issues in keeping with its collective basic research expertise. The committee reviewed the future D&D challenges at major DOE sites (see Chapter 2) and provided recommendations on a research agenda to address these problems (see Chapter 4). The committee also considered the research being sponsored by other U.S. and non-U.S. programs as well as the projects supported in the current EMSP portfolio (see Chapter 3), so that unnecessary duplication of effort can be minimized. Processes for identifying future D&D research needs and for better applying the results of successful research are suggested in Chapter 5. The committee also produced an interim report to advise the EMSP on its fiscal year 2001 proposal call. That report is given in Appendix C.

2
D&D Challenges in the DOE Complex

The Department of Energy (DOE) constructed over 20,000 facilities to support nuclear weapons production and other activities, and approximately 5,000 of these facilities were identified as surplus in 1996 (DOE, 1997a). DOE's strategy for managing its surplus facilities is to turn them over to its Office of Environmental Management (EM) to become part of EM's overall site cleanup program (see Figure 2.1). EM will begin accepting additional facilities in 2002 and expects eventually to receive most of the remaining facilities. Many of these facilities are contaminated with radioactive materials and hazardous materials, such as mercury, asbestos, and lead. These facilities require continued monitoring and maintenance, because deterioration could eventually make them unsafe for workers to enter or increase the risk of releasing contaminants to the environment.

EM has a two-part strategy for dealing with the surplus facilities that it has accepted: deactivation to stabilize each facility and reduce its maintenance costs, and decommissioning when technically and financially appropriate. Facilities that will eventually require deactivation and decommissioning (D&D) include the following:

- production reactors,
- research reactors,
- chemical processing buildings,
- uranium, plutonium, and tritium production facilities, and
- gaseous diffusion plants.

DOE defines deactivation as "a set of integrated and systematic actions [that] render a facility safe and stable until it can be decommissioned" (DOE, 1999c, p. 1). It is a process whereby nuclear materials and chemicals, equipment, and operating systems are placed in a low-risk, low-cost, and mostly passive condition. Included in a facility

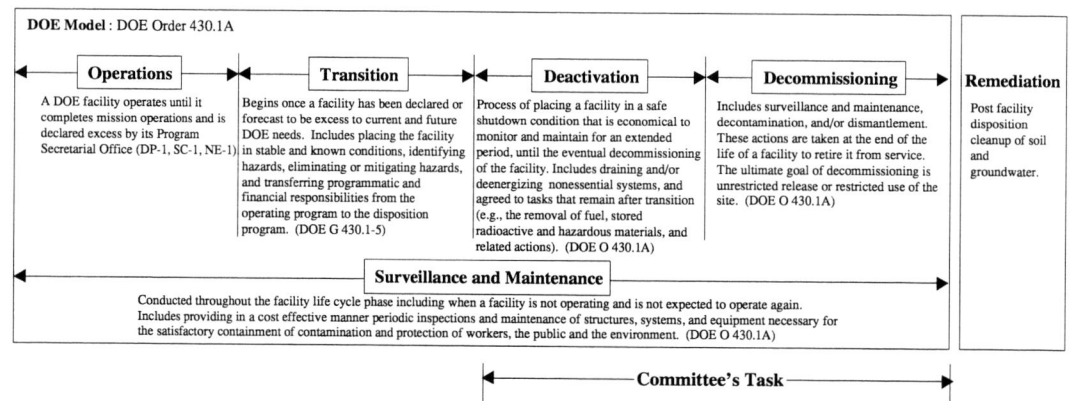

FIGURE 2.1 Transition and disposition phases of a DOE facility's life cycle. Source: DOE.

deactivation, for example, could be the selective shut down of ventilation, electrical, and security systems; removing items from glove boxes and laboratory hoods; and flushing out chemical processing equipment and associated process piping and instrumentation. To ensure safety, surveillance and maintenance of deactivated facilities must continue. However, the cost of these continued activities is expected to be reduced after the deactivation process (Boyd, 2000).

Decommissioning, according to DOE, takes place after deactivation and includes surveillance and maintenance, decontamination, and/or dismantlement (see Figure 2.1). The intended result of decommissioning (referred to as the "end state") is typically decided individually for each facility. In some cases the facility is completely removed and the area is returned to a greenfield with unrestricted future use. In other cases the facility may be reused, and in others sufficient contamination remains to warrant long-term surveillance and institutional control. Post-D&D activities, which are part of DOE's overall site-cleanup program, may include remediating soil or groundwater contamination or long-term surveillance.

D&D work must be done in compliance with all applicable regulations, including those of DOE, the Nuclear Regulatory Commission (USNRC), the Environmental Protection Agency (EPA), and state agencies. In most cases where only DOE has legal authority over its facilities, its regulations are consistent with standards set by USNRC and EPA. DOE is committed to adhering to the ALARA (as low as reasonably achievable) principle for radiation exposure to workers while at the same time minimizing the cost of a D&D project. Deciding the end state of a facility also requires balance between cost and future risks. While these issues are outside the task of this committee, the committee believes that scientific research can provide new ways to reduce radiation exposure and other health risks while reducing cost, and its research recommendations (Chapter 4) are made with these benefits in mind. Better scientific knowledge as input to risk analysis and public decision making will improve all of DOE's site-cleanup activities.

In keeping with the charge to the committee, this chapter first describes many of the DOE facilities that will eventually undergo D&D, with emphasis on those that could be problematic (e.g., expensive or pose potential risk to workers, the public, or the environment) to D&D with currently available (baseline) technologies.[1] This chapter next describes the phases of a D&D project, with emphasis on steps where the committee believes that new technologies could be most useful and where research may provide the greatest benefit. Finally the chapter discusses the objectives ("end states") for facility decommissioning and where research is needed to help support the selection of end states.

Overview of DOE Facilities

Nuclear weapons production in the United States was a complex series of integrated activities executed at multiple sites across the country. These activities can be grouped into eight major processes:

- mining, milling, and refining of uranium;
- isotope separation of uranium, lithium, boron and heavy water;
- fuel and target fabrication for production reactors;
- reactor operations to irradiate fuel and targets to produce nuclear materials;
- chemical separations of plutonium, uranium, and tritium from irradiated fuel and target elements;
- component fabrication of both nuclear and nonnuclear components;
- weapon operations, including assembly, maintenance, modification, and dismantlement of nuclear weapons; and
- research, development, and testing.[2]

Figure 2.2 illustrates these processes and locates the major sites where they were used (DOE, 1997a, p.15), and Table 2.1 lists the processes and facilities in greater detail.

---

[1]Within EM, baseline technologies are currently available and sufficiently established for a contractor to use as the basis for estimating cost and duration of a D&D task.

[2]Nuclear weapons research, development, and testing take place concurrently with the other seven processes. Research and development are mostly complete before component fabrication begins, but testing may continue until a weapons system is retired from the stockpile.

FIGURE 2.2 The United States nuclear weapons complex included facilities that were constructed throughout the country. This figure indicates the location of some of the major facilities and depicts the key production steps. Source: DOE, 1996.

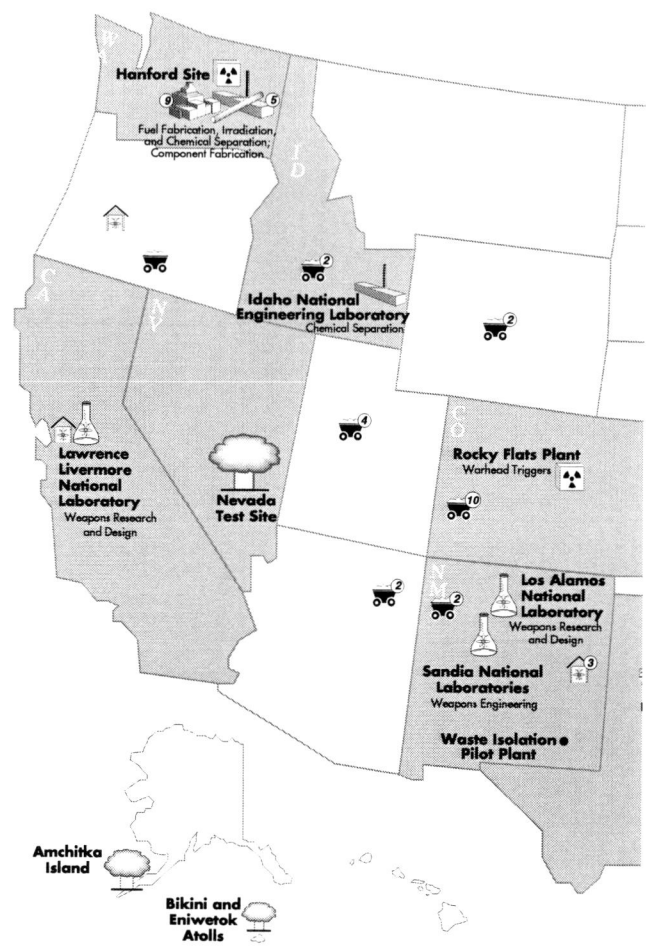

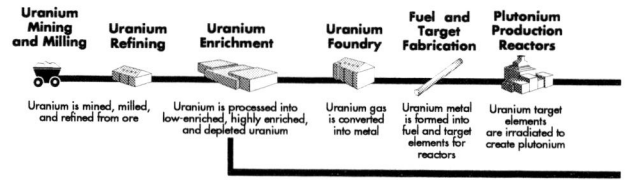

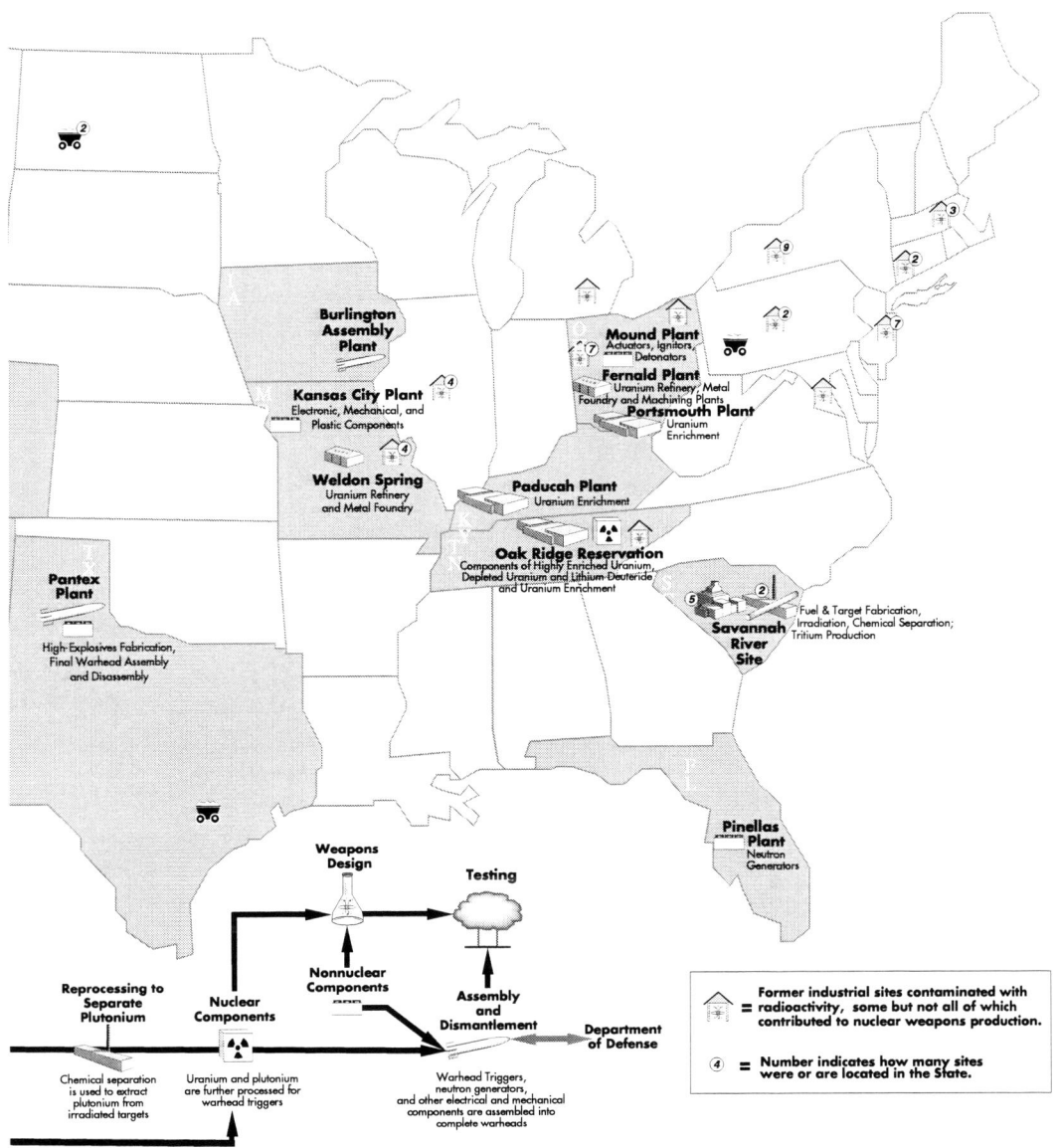

Chapter 2

TABLE 2.1 Weapons Production Processes at the Major DOE Sites

Step	Process	Major Sites
1	Uranium mining, milling, and refining	*Mining and Milling:* Uranium Mill Tailing Remedial Action Project mining and milling sites; other commercial-owned domestic mines; other commercial- and government-owned mills; foreign suppliers. *Ore sampling:* Fernald and Middlesex Refining: Fernald and Weldon Spring (natural, depleted, and enriched uranium reactor fuel and targets); Oak Ridge Y-12 (weapon parts and highly enriched reactor fuel); Oak Ridge K-25, Paducah, and Portsmouth gaseous diffusion plants (production of UF6 feed).
2	Isotope separation	*Uranium:* Oak Ridge K-25; Paducah; and Portsmouth gaseous diffusion plants. *Lithium:* Oak Ridge Y-12 COLEX and ELEX Plants. *Heavy water:* Savannah River Site Heavy Water Plant; Dana Heavy Water Plant.
3	Fuel and target fabrication	*HEU:* Savannah River Site 300 M Area. *Other uranium:* Fernald; Ashtabula; Hanford 300 Area; and Savannah River Site 300 M Area. *Enriched lithium:* Oak Ridge Y-12 and Savannah River Site M Area.
4	Production reactor operations	*Hanford:* B, D, F, H, DR, C, KW, KE, and N reactors. *Savannah River Site:* R, P, K, L, and C reactors.
5	Chemical separations	*Weapons plutonium:* Hanford 200 East and West areas (PUREX, REDOX, T and B plants, 231-Z Plant); Savannah River Site (F Canyon complex). *Uranium recycling:* Hanford (PUREX, UO3 Plant, REDOX, U Plant); Savannah River Site (H Canyon complex); Idaho National Environmental Engineering Laboratory (Idaho Chemical Processing Plant). *Tritium:* Savannah River Site (Tritium Facility 230H series).
6	Weapons component fabrication	*Plutonium:* Rocky Flats; Hanford 234-5 Plutonium Finishing Plant; Los Alamos (TA-21 and TA-55). *Highly enriched and depleted uranium:* Oak Ridge Y-12; Rocky Flats. *Tritium (including recovery and recycling):* Mound; Savannah River Site (Tritium Facility). *Lithium-6 deuteride (including recovery and recycling):* Oak Ridge Y-12. *Plutonium recycling:* Rocky Flats; Los Alamos (TA-55); Hanford Plutonium Finishing Plant.
7	Weapons operations	*Other non-nuclear:* Pantex; Oak Ridge Y-12; Mound; Kansas City; Pinellas. *Assembly and dismantlement:* Sandia; Pantex; Burlington. *Modifications and maintenance:* Pantex; Burlington; Sandia; Clarksville; Medina modification centers.
8	Research, development, and testing	*National laboratories:* Los Alamos; Lawrence Livermore; Sandia (New Mexico and California). *Test sites:* Nevada Test Site; Bikini and Eniwetok Atolls; Christmas and Johnston Islands; Amchitaka Island; Tonopah Test Range; Salton Sea Test Base.

Source: DOE, 1996, 1997a.

In addition to these historical operations, currently operating (the Defense Waste Processing Facility at the Savannah River Site) and future facilities (e.g., at the Hanford Site and the Idaho National Engineering and Environmental Laboratory) to treat highly radioactive wastes that resulted from DOE's production activities must eventually undergo D&D. Together these facilities represent the spectrum of challenges that EM faces in its D&D task.

Uranium Mining, Milling, and Refining

Mining and milling involve extracting uranium ore from the earth and chemically processing it to prepare uranium concentrate (U_3O_8), sometimes called uranium octaoxide or yellowcake. Uranium concentrates were chemically converted to purified forms suitable as feed materials for the next step in the process. Examples of these feed materials are uranium hexafluoride (UF_6) for enrichment at gaseous diffusion plants and uranium tetrafluoride (UF_4) or uranium metal for fuel and target fabrication. Refining, as discussed in this report, also involves the recycling of various production scraps, production residues, and uranium recovered from fuel reprocessing.

Wartime uranium refining was performed by various contractors in several eastern states. After the war, the Atomic Energy Commission (the predecessor of the DOE) built uranium refineries in Fernald, Ohio, and Weldon Spring, Missouri.

Isotope Separation (Enrichment)

Enrichment is the process of separating naturally occurring isotopes of the same element in order to concentrate the isotope of interest. The three elements that have been isotopically enriched in large quantities for use in the nuclear weapons complex are uranium, lithium, and hydrogen.

Uranium Enrichment

The first U.S. uranium enrichment facilities were located in Oak Ridge, Tennessee. Additional enrichment plants were later built in Piketon, Ohio, and Paducah, Kentucky. Uranium enrichment began with natural uranium and resulted in enriched uranium and depleted uranium. Highly enriched uranium (HEU) contains 20 weight percent or more of uranium-235; it was fashioned into weapons components and also used as reactor fuel. Low enriched uranium (LEU), which contains less than 20 weight percent of uranium-235, and natural uranium were used as reactor fuel for plutonium production. Depleted uranium was used in weapon components and as targets for the plutonium-239 production. All the uranium enriched during the Manhattan Project was HEU for weapons components. However, as early as 1950, LEU was used for reactor fuel.

Uranium enrichment has resulted in large amounts of depleted uranium in storage, large surplus facilities, uranium-contaminated scrap metal (from facility dismantlement), and waste contaminated with polychlorinated biphenyls (from electrical equipment used in the enrichment facilities). Uranium, technetium-99, and organic solvent contamination of soils and groundwater have occurred at a number of DOE sites as a result of uranium enrichment.

Lithium Enrichment

Lithium enriched in the lighter lithium-6 isotope was placed in production reactors to produce tritium and was also chemically compounded with deuterium to be used as a component in nuclear weapons. Lithium was enriched at the Y-12 Plant in Oak Ridge, Tennessee, using the column exchange (COLEX) and electric exchange (ELEX) processes. Both lithium enrichment processes used large amounts of mercury, and as a result, mercury is a major feature of the contaminated facilities at Y-12.

Heavy Water Production

Heavy water is used as a source of deuterium for weapons and as a moderator and coolant for nuclear reactors. Natural water contains small amounts of deuterium (0.015 percent), which was concentrated by a combination of hydrogen sulfide-water chemical exchange, water distillation, and electrolytic processes. Heavy water plants were located in Newport, Indiana, and at the Savannah River Site in South Carolina. Routine industrial practices were used to decommission and dismantle these plants, and no long-term environmental challenges are foreseen.

Fuel and Target Fabrication

Fuel and target fabrication consists of the foundry and machine shop operations required to convert uranium feed material, principally metal, into fuel and target elements used in nuclear materials production reactors. Some later production reactors used separate fuel and target elements, while early production reactors used the same elements for both fuel and targets. Uranium ingots were extruded, rolled, drawn, swaged, straightened, and outgassed to produce rods and plates. The rods were machined, ground, cleaned, coated, clad, and assembled into finished fuel.

Reactor fuel and target fabrication was initially carried out by private contractors and at the Hanford, Washington, and the Savannah River, South Carolina, production reactor sites. Within a decade, government-owned plants in Fernald, Ohio, and Weldon Spring, Missouri, took over part of this mission, supplying the fuel manufacturing plants at Hanford and the Savannah River Site.

Chemical conversion of uranium feed to metal and processing of uranium scrap and residue resulted in facilities contaminated with uranium, acids, and solvents. Uranium metallurgy and machining also resulted in facilities becoming contaminated with uranium.

Reactor Operations

Reactor operations include loading and removal of fuel and target elements, reactor maintenance, and the operation of the reactor itself. Early experimental reactors were built in the Chicago area, at Oak Ridge, and at Hanford. Nine full-scale production reactors were located at Hanford, Washington, and five others were built at the Savannah River Site in South Carolina. Figure 2.3 shows the Hanford B reactor, which is currently undergoing D&D for future use as a museum.

Reactor operations created essentially all the radioactive materials used in the DOE complex. Irradiated fuel and targets were highly radioactive. The components of the reactor cores also became highly radioactive over time. The highly radioactive spent fuel and target materials typically went on to chemical separations, although an inventory of unprocessed spent fuel and targets remains in storage. The reactors also required a large number of support facilities that are now contaminated surplus.

FIGURE 2.3 Built in less than one year, Hanford's B reactor was the world's first plutonium production reactor. After decommissioning it will be preserved as a National Historic Landmark. Source: DOE, 1996.

Chemical Separations

Chemical separation is the process of dissolving spent nuclear fuel and targets and isolating and concentrating the plutonium, uranium, and other nuclear materials they contain. This category also includes the reprocessing of spent nuclear fuel to recover, purify, and recycle uranium for reuse in the nuclear weapons programs and the recovery of uranium from high-level waste at Hanford. Three basic chemical separation processes were used on a production scale in the United States: bismuth phosphate, reduction oxidation (REDOX), and plutonium uranium extraction (PUREX). Chemical separation plants were located at Hanford, Washington; the Savannah River Site, South Carolina; and the Idaho National Engineering and Environmental Laboratory.

Chemical separation of spent fuel and target elements produced large volumes of highly radioactive waste (high-level waste) and large quantities of low-level radioactive wastewater, solid low-level waste, and mixed low-level waste. Also included in this category is chemical processing to recover, purify, and recycle plutonium, uranium, tritium, and lithium from retired warheads, from component production scrap and residues, and from the maintenance, recharging, dismantling, and materials recovery conducted separately on individual components.

Component Fabrication

Weapons component fabrication includes the manufacture, assembly, inspection, bench testing, and verification of specialized nuclear and non-nuclear parts and major subassemblies. The major nuclear component fabrication sites were Los Alamos National Laboratory in New Mexico; the Rocky Flats Plant near Boulder, Colorado; the Y-12 Plant in Oak Ridge, Tennessee; and the Plutonium Finishing Plant in Hanford, Washington. Non-nuclear components were manufactured chiefly at the Mound Plant in Miamisburg, Ohio, the Kansas City Plant in Missouri, the Pinellas Plant in Largo, Florida, and the Pantex Plant near Amarillo, Texas.

Like many conventional manufacturing processes, non-nuclear component fabrication activities have resulted in hazardous waste and contamination of facilities by solvents and heavy metals. The manufacture of high explosives for warheads has resulted in facilities and environmental media that are contaminated with explosive materials.

Weapon Operations

Weapons operations include the assembly, maintenance, and dismantlement of nuclear weapons. Assembly is the final process of joining together separately manufactured components and major parts into complete, functional, and certified nuclear warheads for delivery to the Department of Defense. Maintenance includes the modification and upkeep of a nuclear weapon during its life cycle. Dismantlement involves the reduction of retired warheads to a nonfunctional state and the disposition of their component parts. The dismantlement process yields parts containing special nuclear materials, high explosives, hazardous materials, and other components with hazardous and non-hazardous properties. Some parts are returned to the facility where they were originally produced. Other parts are maintained in storage (e.g., plutonium pits) or are dispositioned onsite. Disposition processes include crushing, shredding, burning of main high-explosive charges, and firing of small energetic components. DOE is the steward of the weapon until all components have been stabilized, stored, and disposed.

Weapons operations were chiefly done at the Pantex Plant near Amarillo, Texas; the Iowa Army Ordnance Plant in Burlington, Iowa; Technical Area 2 of Sandia National Laboratory; and the Clarksville, Tennessee, and Medina, Texas, modification centers.

The legacy of contaminated facilities resulting from assembly and maintenance is relatively small compared to the legacy resulting from the other weapons production steps, because all the radioactive materials handled in this process are generally in the form of sealed weapons components.

Research, Development, and Testing (RD&T)

The main U.S. nuclear weapons research and development facilities are the Los Alamos, Lawrence Livermore, and Sandia national laboratories. Nuclear weapons RD&T include the design, development, and testing of nuclear weapons and their effects. Localized RD&T to support specific site missions (such as fuel fabrication) is generally considered in this report to be part of each site's mission. The committee did not address D&D issues at the weapons test sites that are listed in Table 2.1.

Waste Processing

As a part of site cleanup, wastes that were produced by the historical operations described in the preceding sections and that are currently stored at the DOE sites must be treated and conditioned for disposal. Highly radioactive wastes will be treated in facilities that have similar layout and shielding walls as the formerly used chemical separations plants. One such facility that is now in operation is the Defense Waste Processing Facility (DWPF) at the Savannah River Site in South Carolina.

The facility's mission is to vitrify (convert to a glass-like material) about 36 million gallons of high-level liquid wastes. This mission will be completed by 2026. Similar facilities are planned for the Hanford Site and Idaho National Engineering and Environmental Laboratory (INEEL).

Other waste treatment facilities, such as the Advanced Mixed Waste Treatment Project at INEEL, will treat DOE wastes that contain lower levels of radioactivity. Because these modern waste treatment facilities are being designed—and will be operated and shut down—in view of eventual decommissioning, they are not likely to present as great a D&D challenge as the historical facilities.

Greatest Future Challenges for D&D

DOE recognizes that its site closure program postpones the most difficult D&D tasks until well after 2006 and that this schedule allows time for the development of new technologies to perform at least some of these tasks (Hart, 2000). The committee's statement of task asked the committee to identify significant problems that cannot be addressed effectively with current technologies. As it received presentations from DOE and DOE contractors and toured facilities at Hanford, Oak Ridge, and Rocky Flats, the committee first developed a list of underlying reasons why many of the surplus DOE facilities, especially the larger facilities, will present future D&D challenges. The underlying reasons include the following:

- personnel hazards—penetrating radiation, airborne contamination, chemical hazards, and industrial hazards;
- number and size of the facilities and bulk of concrete shielding walls;
- complex, crowded, often retrofitted equipment arrangements;
- lack of knowledge concerning the history of operations and contamination in old facilities;
- difficulty in identifying and quantifying many of the radioactive and chemical contaminants; and
- lack of consistent, complex-wide objectives (end states) for D&D.

In the committee's view, the DOE facilities that manifest these underlying reasons will present the greatest future challenges for D&D operations. These facilities are described in this section. Overall the committee found that

These facilities pose strong safety and economic incentives for developing and using innovative D&D technologies that may be achieved though scientific research. The long time frame for completing D&D (50 years or more) allows for substantive research to be completed and applied.

Radiochemical Separation Facilities

These irradiated fuel and target reprocessing facilities include Savannah River Site's F and H canyons; Hanford's PUREX, T-Plant, B-Plant, U-Plant, and REDOX; and Idaho National Engineering and Environmental Laboratory's Chemical Processing Plant. These are massive concrete structures—Hanford's PUREX plant is approximately 1000 feet long with walls up to four feet thick (see Figure 2.4). The plant interiors are heavily contaminated and have penetrating radiation levels that preclude personnel entry (DOE, 1997a). Processes have used acids, organic solvents, and hazardous chemicals like mercury and chromium compounds. Leaks and spills through years of operation have caused seepage into concrete and expansion joints, and deposits of material in obscure locations are highly probable.

FIGURE 2.4 Chemical separation of plutonium from other radioactive materials was carried out in Hanford's PUREX plant and similar facilities. They were referred to as canyons because thick walls to protect workers from radiation surrounded the long, remotely operated processing corridor.
Source: http://www.fas.org/irp/imint/doe.htm

Gaseous Diffusion Plants

Uranium enrichment plants in Oak Ridge, Tennessee, Portsmouth, Ohio, and Paducah, Kentucky, are the largest of DOE's surplus facilities. When first built, the Oak Ridge K-25 Gaseous Diffusion Plant was one of the largest roofed structures in the world, covering nearly 43 acres. The K-25 process building is one-half-mile long and 1000 feet wide (see Figure 2.5). Operation of the enrichment plants has involved uranium, asbestos, solvents, polychlorinated biphenyls (PCBs), heavy metals, and other toxic substances (DOE, 1996, 1997a). In addition, recycled uranium has introduced fission products like technetium-99 and alpha-emitting isotopes into these buildings. Opportunities for cost reduction in the D&D of gaseous diffusion plants were examined by a previous NRC committee (NRC, 1996). However, that committee's findings may be superseded by new DOE directives on release criteria for radioactively contaminated metals from these plants (*Nuclear Waste News*, 2000).

FIGURE 2.5 The K-25 site was one of several that provided enriched uranium. This site includes almost 400 buildings that have a total of about 14.4 million square feet of floor space. Source: http://www.fas.org/irp/imint/doe.htm

Plutonium Processing Facilities

Plutonium processing facilities include a variety of chemical processing and weapons component fabrication facilities like Hanford's 231-Z Plant and Plutonium Finishing Plant (234-S); Savannah River Site's FB-Line and HB-Line; Los Alamos TA-55; and Rocky Flats Buildings 707, 771, 776, and 777 (see Figure 2.6). Most of these facili-

FIGURE 2.6 Door leading into a plutonium processing room at the Rocky Flats Site is sealed due to very high levels of alpha radiation within the room. There are more than 20 such rooms at the site.
Source: DOE, 1996.

ties used glove boxes or gloved cabinets at negative pressure to safely handle the plutonium.

The dominant D&D hazard arises from small, airborne particles of plutonium that can be inhaled or ingested or assimilated through puncture wounds or other industrial injuries. In addition, neutron exposure is significant due to alpha/neutron reactions from plutonium alpha particles interacting with low-atomic-weight elements like oxygen and fluorine. Furthermore, the penetrating radiation dose in plutonium facilities increases over time due to the ingrowth of americium-241 (from beta decay of plutonium-241). Potential accumulation of plutonium in piping, vessels, and ventilation ducts could be encountered during D&D. In some cases accumulation may be sufficient to raise nuclear criticality issues.[3] A variety of acids, solvents, and other chemicals were used in the facilities.

Glove boxes are often crowded with equipment, and modifications have created areas where there is little room for workers to perform D&D tasks. These areas are also highly contaminated, which leads to slow D&D work, with workers encumbered by protective clothing, including respirators or plastic suits to provide clean air for breathing.

Tritium Processing Facilities

Tritium facilities, although fewer than DOE plutonium and uranium facilities, are D&D challenges. The Savannah River Site has been the United States' primary tritium production facility since the mid-1950s. The Savannah River Site also purifies and loads tritium into weapons components.

Tritium represents a unique radiological hazard because of its volatile forms and ability to undergo isotopic exchange with any hydrogen-containing material (e.g., water, grease) that it contacts. Because it is a very-low-energy beta emitter, it is difficult to detect inside equipment and waste packages. Further, it diffuses into metals and concrete. Hazardous metals like mercury are also present in tritium facilities; for example, mercury diffusion pumps are commonly used. The mercury becomes contaminated with tritium and can be spilled inside the process lines during maintenance work. D&D waste will likely present challenging characterization problems to distinguish between low-level radioactive versus low-level mixed waste.

---

[3]According to DOE, the sudden shutdown of plants that handled plutonium residues and scraps stranded 26 tons of plutonium in intermediate processing steps (DOE, 1996).

The original Savannah River Site tritium processing facility, Building 232-F, has been decommissioned. It began extracting tritium from lithium-6 aluminum target elements irradiated in the five Savannah River Site production reactors in 1955, was replaced in 1958, and sat idle until it was decommissioned between 1994 and 1996. During D&D, surprises were encountered that prompted a search for better detection, bulk sampling, and laboratory analysis methods for characterizing contaminated concrete (Hochel, 2000).

Phases of a D&D Project

The D&D of nuclear facilities is a complex, multi-step process. The DOE model shown in Figure 2.1 describes some of the steps. As an aid to its own fact finding and discussions, the committee has chosen to describe D&D, in a simplified way, as a five-step process that begins with deactivation and ends with formal closing of the project (see Table 2.2):

- a facility deactivation phase following the permanent shutdown of all operations;
- an assessment and decision making phase;
- a planning phase for the development of the D&D plan;
- the physical decontamination and decommissioning operations phase; and
- a closeout phase during which the end-state requirements have been verified and the facility can be released for its intended end use or long-term monitoring.

Table 2.2 illustrates these five phases and some of the key activities that occur in each phase. Table 2.2 also notes areas where the committee identified opportunities to significantly reduce cost and enhance the safety of workers, the public, and the environment. These areas are characterization, decontamination, robotics, and scientific input into determining the facility end state. In Chapter 4, the committee provides research recommendations for each of these areas.

Facility End States

For a facility undergoing D&D, the project's objective is a major determinant of the cost, schedule, and amount of work required (which affects risk to workers). This objective is often referred to as the facility

TABLE 2.2 Phases, Activities, and Opportunities of a D&D Project

Phase 1 Facility Deactivation (Surveillance and Maintenance)	Phase 2 Assessment and Decision Making
Shut down Systems	Establish D&D Objectives: -Facility End Use & End State
Regulatory Compliance	-Regulatory Requirements -Financial Considerations

Public & Community
  Communications Program
	Identify Technology and
	  Science Needs

CHARACTERIZATION -Contamination Control and Surveillance	Perform Risk Assessments
- Safeguards and Security	Obtain External Science Reviews
DECONTAMINATION -Personnel access	Evaluate Alternatives
-Contamination control	Encourage Public Involvement
ROBOTICS -Inspection	Select D&D Alternative
-Maintenance -Monitoring	CHARACTERIZATION -Conduct Surveys -Obtain & Analyze Samples

end state. End states for individual facilities or buildings on a site are generally part of the overall cleanup agreement for the site—the site "end state." Even if a facility is completely removed, there are post-D&D remediation activities as shown in Figure 2.1. End states are negotiated by the DOE working with state and federal regulators and local stakeholders, and for most facilities and buildings they have not been defined. Viewed from this perspective it is not surprising or undesirable that there is no complex-wide, consistent definition of end states.

Previous NRC study committees have examined the question of end states for DOE cleanup activities. One of these reports discussed end states in terms of alternative scenarios and accordingly defined end state as the "the final product of a waste processing, remediation, or management scenario characterized well enough in terms of chemical, physical, and radioactive attributes to allow details of scenarios to be

Phase 3 D&D Plan Development	Phase 4 Decommissioning Operations	Phase 5 Facility Closeout
Select Acquisition Strategy	Equipment & Structure Removal	Post Decommissioning
Schedule Baseline - Cost Baseline	Monitoring of Workers	Regulatory Reviews

Initiate New Technology
  Development
	Area Monitoring and Interim
	  Release of Areas
		Release Facility for Reuse, Monitoring or
		Long-term Stewardship

Use CHARACTERIZATION
  DATA to Complete D&D Plans
      -D&D Sequence
      -Permits & Licenses
      -Man-Rem Estimates
      -Waste Quantities Est.
	Implement D&D Safety
	  Program
	
	Continue Public
	  Communications Program
	
	DECONTAMINATION of
	  Components, Systems, and
		Issue Lessons Learned
		
		CHARACTERIZATION to Verify
		END STATE
		  -Independent Contractor Surveys
		  -Issue Post Decommissioning Report

Select CHARACTERIZATION and Monitoring Methodology	Structures	

Establish END STATE
  Requirements
	DECONTAMINATION &
	CHARACTERIZATION of
	  Materials	

Select ROBOTIC Systems	CHARACTERIZATION and Management of Waste	

specified" (NRC, 1999a, p. 2). Another study viewed end states as guides to site disposition decisions and found that "end states appear at present to be emerging as the de facto result of multiple interim actions" (NRC, 2000b, p. 21). Decisions could in principle be revised as new information became available or as ideas change. Focusing more narrowly on end states for facility D&D, the previous study of the D&D focus area (see Chapter 3) found that "[w]ithout defining 'how clean is clean enough' the necessary technology, cost, and schedule for a D&D project cannot be determined" (NRC, 1998a, p. 3).

In its fact finding the committee encountered a variety of site agreements and guidelines for D&D end states. For most facilities at Rocky Flats the end state is complete removal of the building. Re-industrialization, including decontamination and reuse of many buildings, is planned for much of the Oak Ridge Site and parts of the Hanford Site.

The Canyon Disposition Initiative at Hanford is considering an option of partly decontaminating a separations building, filling the building with low-level waste, and mounding earth over the building. In the latter two examples, the agreed upon end state will determine the degree to which the facilities must be decontaminated.

While establishing end states is clearly a matter of public negotiation and decision making, the committee believes that a better scientific underpinning for the selection of facility end states is needed. The committee believes that there are two major areas in which the scientific basis for evaluating the risk to public and worker health and safety of various end states is inadequate.

The first area is the understanding of the actual health risk of residual levels of radioactive and hazardous materials (MacLachlan, 2000). The EMSP has acknowledged that "[t]here is scientific uncertainty regarding a safe level of risk to human health and the environment for the end state of the Department's cleanup effort" (DOE, 2000g, p. 7). In cooperation with the DOE Office of Science, the EMSP recently issued a research solicitation on low-dose radiation effects, which included four specific areas aimed at developing a better understanding of risks. The Nuclear Regulatory Commission (USNRC) has recently asked for advice on the technical bases for criteria to control release of slightly contaminated solids from USNRC-licensed facilities.[4] Data on radiation exposure to D&D workers will be valuable for future epidemiological studies.

The other area in which there is inadequate scientific basis for evaluating the health risk of end states is the fate and behavior of contaminants in D&D facilities. The EMSP recognizes that "[a]ccurate risk analysis requires thorough knowledge of contaminant characteristics. . . . Research is needed to define transport dynamics of toxic chemicals and radionuclides" (DOE, 2000g, p. 7). The committee's recommendations for research in this area are given in Chapter 4.

---

[4]Board on Energy and Environmental Systems, "Alternatives for Controlling the Release of Solid Materials from Nuclear Regulatory Commission-Licensed Facilities." In progress.

3

Current D&D Science and Technology Programs

This chapter provides an overview of U.S. and international science and technology programs directed at nuclear facility deactivation and decommissioning (D&D). It is intended to provide information for readers not familiar with D&D programs and particularly researchers who may be interested in submitting a proposal to the Department of Energy's Environmental Management Science Program (DOE-EMSP).

The chapter also responds to the third part of the committee's charge "In recommending specific areas of research, the committee should take into account, where possible, the agendas of other D&D-related research programs." The committee found that the EMSP is unique in its targeting of research funding toward DOE's D&D challenges. However, some programs funded by other U.S. agencies and industries are relevant to D&D, and they may provide good leads for new or improved D&D approaches. In addition, there are opportunities for international collaborations in the area of nuclear facility D&D.

U.S. Department of Energy Programs

The DOE's site cleanup program is one of the largest environmental cleanup efforts in world history. The program is estimated to cost over $100 billion, with some estimates exceeding $200 billion, and to continue for decades (DOE, 1999c, 2000d). To deal with this task DOE established its Office of Environmental Management (EM) in November 1989. The mission of the EM program is to bring DOE sites into compliance with all applicable regulations while minimizing risks to the environment and human health and safety posed by cleanup operations.

EM's current approach to site cleanup is to make the maximum use of existing and proven technologies to achieve as much cleanup as practicable in the shortest time. The goal of EM's Paths to Closure, initially stated in 1996, was to clean up 90 percent of its sites within the next ten years (DOE, 1998a). DOE's recent *Status Report on Paths to Closure* reiterates the goal to "close as many sites or portions of sites as possible by 2006" (DOE, 2000d, p. 2). The status report notes significant accomplishments and progress at most DOE sites. Presentations to the committee as well as its site tours provided an overview of completed D&D work and work in progress.

The *Paths to Closure* approach reduces DOE's fixed costs of maintaining infrastructure of a site as rapidly as possible and provides evidence of progress in the cleanup program. DOE recognizes that the more intractable and expensive cleanup tasks are postponed until well in the future. This provides both the time and incentive to improve current technologies or develop new ways to address these more difficult challenges. Specifically for facility D&D, DOE estimates that about 80 percent of the costs will be incurred after 2006 (Hart, 2000).

The Office of Science and Technology (OST) is the EM division charged with developing or finding new technologies to assist the cleanup mission. Research and development investments by OST have the objectives to

- meet the high-priority needs identified by the cleanup project managers,
- reduce the cost of EM's costliest cleanup projects,
- reduce the technological and programmatic risk of completing major cleanup projects on time and within budget, and
- accelerate and increase technology deployments (DOE, 2000e).

To identify technology needs, OST has formed a site technology coordinating group (STCG) at each major DOE-EM site to interact with local contractor personnel and others. Each group includes senior managers from the site DOE office, site contractors, and national laboratory personnel. The STCGs are responsible for developing and prioritizing a list of site problems and technology needs based on environmental management issues relevant to a specific site. Each STCG evaluates and prioritizes technology needs according to a set of criteria established by the STCG (NRC, 1999e). The committee's suggestions for improving the process for identifying site needs are given in Chapter 5.

OST has also created a gate system as a means of managing its technology development program (Boyd, 2000). Projects originating from basic research begin at gate 1, and if successful, continue through applied research, development, demonstration, and eventually, deployment. This gate concept is shown in Figure 3.1, which also depicts the mixture of personnel involved at each gate, beginning with basic researchers and ending primarily with site contractors and industries.

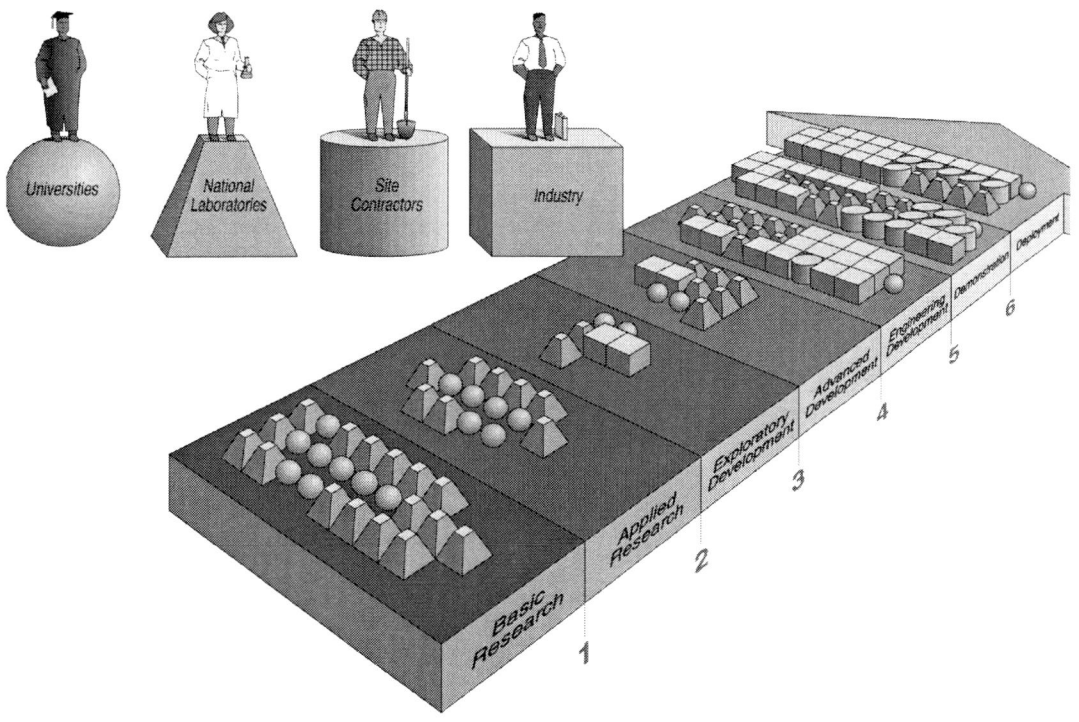

FIGURE 3.1 The DOE Office of Science and Technology uses a gate system that begins with basic research and continues through deployment of new technologies.
Source: DOE.

There are two major programs within OST that are designed to bring new technologies through the gate system. The Environmental Management Science Program (EMSP) is intended to provide mission-driven research, which has elements of both basic (gate 1) and applied (gate 2) research. The OST focus areas are primarily responsible for developing, demonstrating, and deploying new technologies (gates 3 through 6). The EMSP and focus areas are described in the next two sections of this report.

The Environmental Management Science Program

The EMSP is a collaborative partnership among the DOE Office of Science, OST, and the Idaho Operations Office[1] to sponsor basic environmental and waste-management-related research (see Figure 3.2). Research supported by the EMSP is expected to have relevance to EM's most challenging technical problems. Results are expected to lead to new knowledge and technologies that reduce the costs, schedule, and risks associated with DOE's site cleanup program (DOE, 1998a, 2000d).

---

[1] The Idaho Operations Office administers EMSP grants that have been awarded.

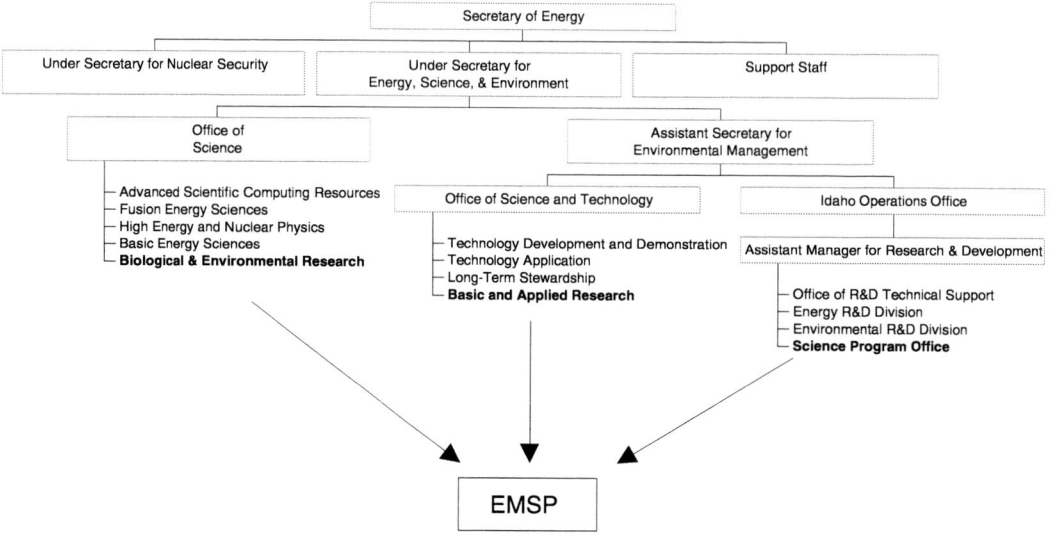

FIGURE 3.2 Department of Energy organizational chart showing the offices that comprise the Environmental Management Science Program.

The EMSP was established in response to a mandate from Congress in the fiscal year 1996 Energy and Water Development Appropriations Act. Congress directed DOE to "provide sufficient attention and resources to longer-term basic science research which needs to be done to ultimately reduce cleanup costs, . . . develop a program that takes advantage of laboratory and university expertise, and . . . seek new and innovative cleanup methods to replace current conventional approaches which are often costly and ineffective" (DOE, 2000g, p.1).

The key EMSP activities include identifying and prioritizing research needs, developing proposal solicitations, selecting research projects, managing its research portfolio, funding the selected research, and integrating that research into technology development projects. Figure 3.3 depicts the overall process the EMSP uses to select research projects for funding. EMSP research needs are identified through a variety of inputs, including the focus areas, the SCTGs, and site cleanup managers. External evaluations, such as this report and others from the National Academies, are also used by the EMSP for writing its solicitations and evaluating proposals.

The EMSP's annual solicitations usually target needs in one or two of OST's focus areas. For example, this committee's interim report (see Appendix C) and the present report were both requested to assist the EMSP in its Fiscal Year 2001 call for proposals in the area of D&D. Successful proposals are usually awarded for three years, so that the selection/award process for a given focus area repeats in approximately three-year cycles. If a large number of proposals are expected, scientists seeking awards are encouraged to submit a brief pre-proposal for evalu-

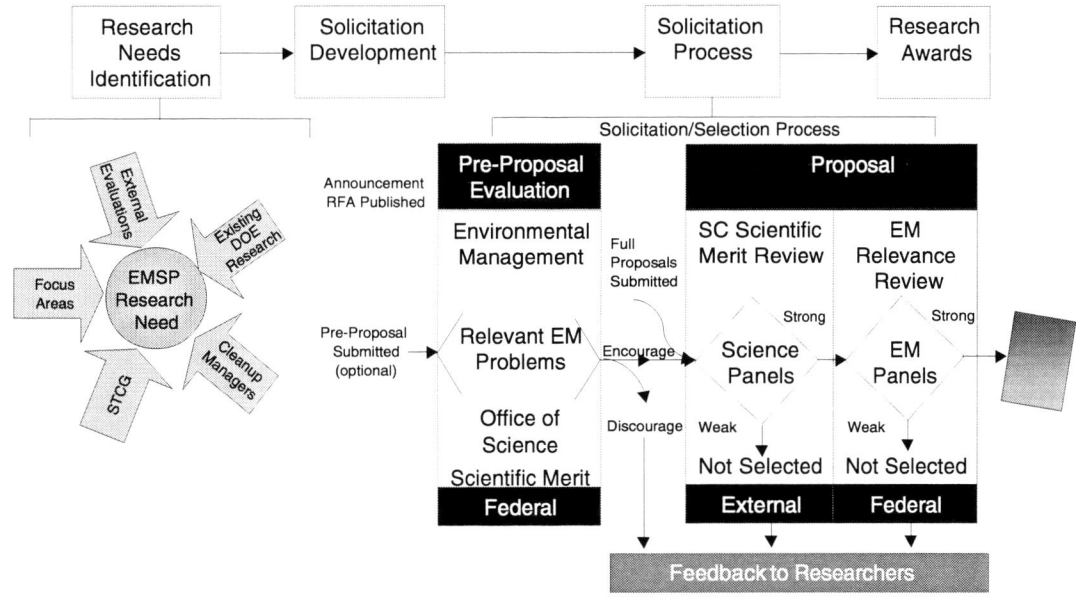

FIGURE 3.3 Selection of research proposals by the EMSP includes review for scientific merit and relevance to EM needs. Research needs are identified from a number of inputs. Source: DOE.

ation. Promising research proposals can therefore be identified early in the process.

Full proposals undergo a two-step review. The first is for scientific merit, and it is conducted by an independent peer review panel under the auspices of the DOE Office of Science. The second is for relevance to EM needs. The overall process is thus intended to select proposals that are both scientifically sound and that stand a good chance of leading to improved technologies for application in the EM program. Table 3.1 summarizes EMSP proposals in the D&D area that were selected for funding in the previous two calls for D&D proposals. For the 2001 call, 54 D&D-related proposals were submitted.

In view of the EMSP's congressional charter and the future D&D challenges described in Chapter 2, the committee recommends that:

The EMSP should focus on long-range basic research targeted on broad (site wide) or major (essential to one or a few sites) D&D needs.

Research projects should address significant long-term problems to advance the state of knowledge well beyond the next decade. This approach maintains the EMSP long-term mission. Nevertheless, opportunities for research that provides high potential payoff in addressing urgent near-term needs may arise. As a practical matter, the EMSP may well encounter a range of research opportunities that span short- and long-term needs as well as provide for contingent approaches for D&D.

Chapter 3

TABLE 3.1 Environmental Management Science Program Projects

Project Title	Funding
Synthesis of New Water-Soluble Metal-Binding Polymers: Combinatorial Chemistry Approach	$1,120,000
Atmospheric-Pressure Plasma Cleaning of Contaminated Surfaces	$1,212,001
Advanced Sensing and Control Techniques to Facilitate Semi-Autonomous Decommissioning	$870,733
In-Situ Spectro-Electrochemical Studies of Radionuclide Contaminated Surface Films on Metals and the Mechanism of Their Formation and Dissolution	$1,005,000
Modeling of Diffusion of Plutonium in Other Metals and of Gaseous Species in Plutonium-Based Systems	$435,000
Development of Monitoring and Diagnostic Methods for Robots Used in Remediation of Waste Sites	$402,782
Removal of Radioactive Cations and Anions from Polluted Water Using Ligand-Modified Colloid-Enhanced Ultrafiltration	$538,997
Waste Volume Reduction Using Surface Characterization and Decontamination by Laser Ablation	$789,999
Optimization of Thermochemical, Kinetic, and Electrochemical Factors Governing Partitioning of Radionuclides: Melt Decontamination of Radioactively Contaminated Stainless Steel	$1,200,000
Micelle Formation and Surface Interactions in Supercritical CO_2: Fundamental Studies for the Extraction of Actinides from Contaminated Surfaces	$960,000
Decontamination of Radionuclides from Concrete During and After Thermal Treatment	$815,720
"Green" Biopolymers for Improved Decontamination of Metals from Surfaces: Sorptive Characterization and Costing Properties	$900,000
Improved Decontamination: Interfacial, Transport, and Chemical Properties of Aqueous Surfactant Cleaners	$570,000
Microbially Promoted Solubilization of Steel Corrosion Products and Fate of Associated Actinides	$1,406,000
Mechanisms of Radionuclide-Hydroxycarboxylic Acid Interactions for Decontamination of Metallic Surfaces	$1,150,000
Contaminant-Organic Complexes, Their Structure and Energetics in Surface Decontamination	$1,241,950
Decontamination and Decommissioning of PCB Sites at DOE: Extraction, Electrokinetics, and Hydrothermal Oxidation	$979,808
Metal Ion Analysis Using Near-Infrared Dyes and the "Laboratory-on-a-Chip"	$470,770
Development of Novel, Simple Multianalyte Sensors for Remote Environmental Analysis	$650,000
Real-Time Identification and Characterization of Asbestos and Concrete Materials with Radioactive Contamination	$599,687
Three-Dimensional Positron-Sensitive Germanium Detectors	$750,000
Atmospheric-Pressure Plasma Cleaning of Contaminated Surfaces	n/a

The D&D Focus Area

OST has organized five focus areas to ensure that its efforts efficiently address EM's most urgent technology needs. These include:

- deactivation and decommissioning,
- high-level waste tanks,
- subsurface contaminants,

- transuranic and mixed waste, and
- nuclear materials.

The D&D focus area (DDFA) was established by OST to ensure that adequate technologies are available to support EM's D&D task. In addition to the buildings themselves, the task will include decontamination of the metal and concrete within those buildings and disposal of some 180,000 metric tons of scrap metal and over 27 million tons of concrete.

Most presentations to the committee during its site visits expressed the view that for the D&D task, unlike tasks addressed by the other focus areas, sufficient baseline technologies exist—that there are no substantial technology gaps. However, EM also understands that many baseline D&D technologies are labor intensive, time consuming, expensive, and involve major risks of worker exposure to radioactive and other hazardous materials. In addition, many baseline technologies generate secondary wastes (waste generated by the cleanup operation itself) in relatively large volumes. Some of these secondary wastes are difficult to handle, process, and dispose (NRC, 1998b, 1999b).

The DDFA is charged with providing new technologies that are significantly better than the available baseline technologies[2] in terms of cost, speed, safety, or waste reduction. It therefore develops, demonstrates, and seeks to deploy new technologies that meet one or more of these criteria. A key function for the DDFA is demonstration of promising new technologies to end users—the contractors who actually perform the D&D work. The DDFA conducts technology demonstrations in DOE facilities at a scale and test duration that is convincing to potential end users. Such technology demonstrations comprise the DDFA's large-scale technology demonstration and deployment program. Those programs that have been completed or are still in progress are listed in Table 3.2. The DDFA counts over 300 successful new technology deployments as a result of the program (DOE, 2000f).

The DDFA was reviewed by a previous NRC committee (NRC, 1998a), and the present committee has not directed comments or recommendations to the DDFA. However, because the role of the DDFA is to develop and deploy new technologies, it is clearly necessary that the DDFA pick-up and sponsor promising research results from the EMSP. Suggested ways to improve the handoff, sponsorship, and maturing of EMSP research through OST's gate system are discussed in Chapter 5.

---

[2] Within EM, baseline technologies are currently available and sufficiently established for a contractor to use as the basis for estimating cost and duration of a D&D task.

Chapter 3

TABLE 3.2 Large-Scale Demonstration and Deployment Projects

Current Projects	Completed Projects
• Los Alamos National Laboratory transuranic waste	• Chicago Pile 5 (CP-5) reactor at Argonne National Laboratory-East
• Mound tritium facilities	• Hanford C reactor
• Idaho National Engineering and Environmental Laboratory fuel storage canals and underwater and underground facilities	• Fernald Plant-1
	• Savannah River 321-M fuel fabrication facility

U.S. Programs Outside the DOE

The committee found no other U.S. programs comparable to the EMSP in funding scientific research aimed at the D&D of nuclear facilities. The committee did find a number of programs, however, that are of significance to the D&D task. These are listed in Table 3.3. The Electric Power Research Institute, for example, has a decommissioning technology and planning program. Both the Environmental Protection Agency and the Nuclear Regulatory Commission have regulatory responsibilities that pertain to facility D&D. Research conducted by military organizations is mission-directed, but in a broader sense than that of the EMSP. Some military research (e.g., chemistry, materials science, and intelligent systems) may be applicable to D&D. The National Science Foundation (NSF) is the largest funding sponsor of basic research (e.g., biology, chemistry, materials science, physics) that might find application to D&D problems.[3] A previous NRC study provides a 20-page table of programs in other federal agencies that in the general nature of basic research could be relevant to D&D (NRC, 2000a).

International D&D Science and Technology Progams

Since 1979 the European Commission (EC) has been engaged in scientific research and technology development for nuclear facility D&D.

---

[3]The NSF maintains a web site at http://www.nsf.gov. Readers interested in pursuing international collaborations (see next section) can find information at <http://www.nsf.gov/sbe/int/start.htm>.

TABLE 3.3 Research Programs Outside of the U.S. Department of Energy

PROGRAM **WEBSITE**
Description and Funding

ELECTRIC POWER RESEARCH INSTITUTE (EPRI) Available at: www.epri.com

The goal of the EPRI decommissioning technology program is to assist utilities minimize the cost of decommissioning through enhanced planning, determining optimum financial fund set-aside, applying lessons learned by other utilities with retired plants, and use of advanced technology. Recently expanded with the participation of the five U.S. utilities that have shutdown nuclear plants but no operating nuclear units, this program now has a technical steering committee open to domestic utilities with an interest in decommissioning and international utilities that join the program.

Another goal is to facilitate transfer of relevant technical developments from the programs of the Department of Energy, especially from the Federal Energy Technology Center (FETC), and from international organizations.

EPRI is currently evaluating the USNRC's D&D and the DOE's RESRAD codes to understand the differences and the implications of each code in modeling and meeting the 25 mrem/yr limit.

Funding is $1M per year; less than $100,000 per year goes to universities.

NUCLEAR ENERGY INSTITUTE (NEI) Available at: www.nei.org

Traditionally, the NEI does not fund science-based research, as it is primarily concerned with regulatory development and implementation issues. However, they fund research indirectly by funneling funds into EPRI's efforts.

U.S. NUCLEAR REGULATORY COMMISSION (USNRC) Available at: www.nrc.gov

Nuclear Facilities Decommissioning Program

USNRC regulatory and oversight activities include decommissioning, which involves safely removing a facility from service and reducing residual radioactivity to a level that permits the property to be released. This action is to be taken by a licensee before termination of the license. Some power reactor licensees recently have decided to shut down their facilities prematurely, before the expiration of the current operating licenses (e.g., Haddam Heck, Maine Yankee, Zion, etc.). These unexpected shutdowns have resulted in additional staff efforts in the areas of decommissioning inspections and in the licensing area to process license amendments and exemptions reducing regulatory requirements to correspond to the reduced risk posed by the permanently shutdown plants. In some cases, non-licensed facilities may also be required to reduce or stabilize contamination before sites are released. This activity comprises the USNRC's integrated regulation of the decontamination and decommissioning of facilities and sites associated with NRC-licensed activities, including associated research, rule-making efforts, and the technical interface with the Environmental Protection Agency (EPA) to resolve issues of mutual interest in accordance with the March 1992 General Memorandum of Understanding.

U.S. ENVIRONMENTAL PROTECTION AGENCY (EPA) Available at: www.epa.gov

Environmental Technology Verification Program (ETV)

The ETV, a new program, was instituted to verify the performance of innovative technical solutions to problems that threaten human health or the environment. ETV was created to substantially accelerate the entrance of new environmental technologies into the domestic and international marketplace. ETV verifies commercial-ready, private-sector technologies through 12 pilots.

The "Waste Research Strategy" covers research necessary to support both the proper management of solid and hazardous wastes and the effective remediation of contaminated waste sites. This research includes improving the assessment of existing environmental risks, as well as developing more cost-effective ways to reduce those risks.

Pollution Prevention Research Strategy

One of EPA's strategic goals is to prevent pollution and reduce risk in communities, homes, workplaces, and ecosystems. This goal must be based in large part on the application of the best available science and technology associated with a preventive approach. The Office of Research and Development's (ORD's) Pollution Prevention Research Strategy describes the agency's research and development program in pollution prevention for the next five years. It is designed around the vision that scientifically based pollution prevention research and development products will be used routinely for improved environmental decision making on high-risk human health and environmental problems.

continued

TABLE 3.3 Continued

PROGRAM **WEBSITE**
Description and Funding

National Center for Clean Industrial and Treatment Technologies (CenCITT)

CenCITT is a research consortium dedicated to advancing science, engineering, and pollution prevention. It was established through a base grant from EPA's Centers Program. Its founding members included Michigan Technological University (MTU), the University of Wisconsin-Madison (UW), and the University of Minnesota-Twin Cities (UM). Since its establishment CenCITT has initiated 57 projects involving 51 principal investigators, 57 companies, 33 governmental and other organizations, and well over 100 students. Targeted industry sectors have included chemical processing, metals, manufacturing, energy, forest products, and others. Participating disciplines have included environmental, chemical, civil, mechanical, metallurgical, and geological engineering, as well as chemistry, biology, social science, business, and forestry.

National Center for Environmental Assessment (NCEA)

Science to Improve Risk Assessment

ORD/NCEA continues to be a national leader in the field of risk analysis of human health and ecological effects, and this effort will continue to serve as a catalyst for NCEA's research program. Drawing on its experience in performing human health and ecological risk assessments, NCEA performs and supports scientific activities designed to improve future risk assessments. A major goal is to perform the research necessary to develop an accessible, seamless, common methodology for combined human health and ecological risk assessments, so that decision makers at all levels can have the integrated view of risk needed to make sound decisions.

National Risk Management Research Laboratory (NRMRL)

The NRMRL conducts research into ways to prevent and reduce risks from pollution that threaten human health and the environment. The laboratory investigates methods and their cost-effectiveness for prevention and control of pollution to air, land, water, and subsurface resources; protection of water quality in public water systems; remediation of contaminated sites, sediments, and ground water; prevention and control of indoor air pollution; and restoration of ecosystems.

National Health and Environmental Effects Research Laboratory (NHEERL)

NHEERL is EPA's focal point for scientific research on the effects of contaminants and environmental stressors on human health and ecosystem integrity. Its research helps the agency identify and understand the processes that cause pollution and helps the agency evaluate risks that pollution poses to humans and ecosystems. The impact of NHEERL's efforts can be felt beyond the agency as well, enabling state and local governments to implement more effective environmental goals, and informing international governments and organizations on issues of environmental importance.

U.S. DEPARTMENT OF ARMY—ARMY RESEARCH OFFICE Available at: www.aro.army.mil

Chemical Sciences Division:

Analytical Chemistry

The detection and identification of chemical agents, explosives, and chemical contaminants are of interest to the Army on the battlefield and for environmental applications. Research areas include novel detection and identification schemes, small lightweight sensors, multidimensional analytical techniques, and predictive and interpretive models. Proposals in this area will only be considered for basic research that is connected with specific Army research programs.

Polymer

Polymeric materials are of critical importance to the Army for soldier protection and materiel. This program supports molecular-level polymer research leading to new polymeric materials that are chemical resistant, have selective permeability, are resistant to environmental changes, and have reduced weight with increased strength. Research areas of current interest include synthesis of new polymers, molecular property studies, computational modeling, self assembly, molecular architecture, and dendritic molecules.

Surface and Interfacial Chemistry

This program supports research on the decomposition of hazardous molecules on well-characterized surfaces and in organized media (e.g. micelles, microemulsions, vesicles, and monolayer films) at liquid-liquid and liquid-solid interfaces. The

TABLE 3.3 Continued

PROGRAM   **WEBSITE**
Description and Funding

---

development of new experimental probes of these reactions is also of interest. The most important species are organo-phosphorus, -sulfur, and –nitrogen molecules, and reactions of organic functional groups on surfaces and in these organized media. The principal reactions of interest are hydrolysis and oxidation, and catalysis is a strongly desired goal of these studies; however, new concepts are encouraged. We also encourage studies of erosion of metal and other surfaces by combustion gases.

Theoretical Chemistry

Army requirements for insensitive munitions, for propellants and explosives with greater energy density, for the control of propellant burning rates, and for controlled energy release from explosives provide a continuing interest in a variety of theoretical explorations. Theoretical investigations may provide predictive capabilities relevant to the properties and behavior of a wide spectrum of energetic materials and their prototypes, more specifically studies of energy transfer mechanisms in condensed phases, the prediction of molecular reactivities, the investigation of heterogeneous reactions, and the prediction of reaction pathways. Theoretical understanding of atoms, molecules, and clusters on surfaces may provide the basis for rational design of catalysts.

Materials Science Division

Physical Behavior of Materials

The program of Physical Behavior of Materials seeks research directed at providing an improved understanding of the fundamental mechanisms and key materials and processing variables that determine the electronic, magnetic, and optical (EMO) properties of materials and affect the reliability of EMO devices. Emphasis is on research that will facilitate the nanostructuring of materials to realize the materials-by-design concept where new and unique materials are constructed on the atomic scale with application-specific properties. This includes research on understanding the underlying thermodynamic and kinetic principles that control the evolution of microstructures, understanding the mechanisms whereby the microstructure affects the physical properties of materials, and developing insight and methodologies for the beneficial utilization and manipulation of defects and microstructure to improve material performance.

Mathematical and Computer Sciences Division

Applied Analysis

The Applied Analysis Program supports Army needs in mathematical modeling and analysis for advanced solid materials, soil and granular materials, fluid flow including reactive flow, photonic bandgap materials, nonlinear dynamics and inverse scattering.

Advanced Solid Materials

The Advanced Solid Materials Program supports mathematical research oriented toward optimizing properties or performance characteristics of highly nonlinear materials, including advanced composites for armor and "smart" materials for sensors. Lightweight, high-strength structural components, including advanced composites, contribute to attaining mobility and protection requirements for U.S. forces (as well as to the fuel efficiency and safety of the U.S. automobile fleet). Advanced composites are challenging to analyze and design because of the presence of many interacting length scales. Smart materials are the functional ingredients of actuators, sensors, and transducers.

Systems and Control

The Systems and Control Program is concerned with modeling, analysis, and design of complex real-time systems, especially as they relate to Army problems in distributed command, control, and communications and in guidance and control of complex semi-automated and automated systems. The program invests in fundamental control theory, intelligent systems, and design and control of smart structures.

Intelligent Systems

Given advances in technology, environmental factors and goals, an intelligent system configures assets to achieve goals or to replan objectives in a fault-tolerant fashion either autonomously or for intelligence augmentation of human-centered systems.

Design and Control of Smart Structures

The Army is interested in developing an analysis capability that includes a combination of mathematical theories of design, control, analysis, and visualization that would aid in the search for an optimal or near-optimal design of smart and adaptive structures.

continued

TABLE 3.3 Continued

PROGRAM	WEBSITE
Description and Funding	

Atomic, Molecular, and Optical Physics Program

    This area includes matter waves and atom optics; nonlinear atomic and molecular processes for sensor protection and optical processing; the development and exploitation of plasmas for toxic gas destruction, pollution reduction, and materials processing; hybrid optical systems; coherent source generation; coherent or ballistic imaging through turbid and scattering media; and atom optics for novel sensors and lithography.

U.S. AIR FORCE OFFICE OF SCIENTIFIC RESEARCH (AFOSR)　　　　　　　　　　　　Available at: www.afosr.af.mil

Human-Machine Interface

    A basic research program investigating the key enabling technologies necessary to enhance human-machine interfaces. Emphasis is on human factors related to the design of novel interfaces. New measures of individual cognitive workload will be studied for use in monitoring performance in response to stress and fatigue and for establishing benchmarks for alternative interface systems.

Chemistry and Life Sciences

    A wide range of fundamental chemistry and life sciences research is supported to provide the Air Force with novel options to increase performance and operational flexibility. The chemistry effort in the directorate supports the structural materials activities in the Directorate of Aerospace and Materials Sciences to make an integrated AFOSR structural materials program.

Polymer Chemistry

    The goal of this research area is to gain a better understanding of the influence of chemical structures and processing conditions on the properties and behaviors of polymeric and organic materials. This understanding will lead to development of advanced polymeric materials for Air Force applications. Our approach is to study the chemistry and physics of these materials through synthesis, processing, and characterization. This area addresses both functional properties and properties pertinent to structural applications. Materials with these properties will provide capabilities for future Air Force systems to achieve global awareness, global mobility, and space operations as envisioned in New World Vistas.

Surface Science

    Surface science supports basic research in chemistry on the interface, reactivity, and analysis of surfaces and thin films. Our goal is to improve our understanding of surface processes involved in these areas. Research in the chemistry and morphology at interfaces will lead to a better understanding of the mechanisms involved in those surface processes, which in turn will lead to more effective modification and control of surface relationships.

Toxic Biological Interactions

    Air Force operations utilize physical and chemical agents that may interact with biological tissue and be potentially harmful to military and civilian personnel, to the surrounding populace, and to the environment. The agents include non-ionizing radiant energies (radio frequency radiation, microwaves, and laser light), heavy metals (chromium and cadmium), and various chemicals that constitute fuels, propellants, and lubricants of interest to the Air Force. Exposure to these agents may result directly from their use during Air Force operations and maintenance and, in the case of chemicals, may also occur indirectly as a result of leaky storage containers, for example, that contaminate waste streams, groundwater, and soil. To protect humans and maintain safe working environments, the Air Force supports basic research that endeavors to understand how these agents may interact with biological systems at the subcellular and molecular levels to produce toxic effects.

Computational Mathematics

    This program aims to develop improved mathematical methods and algorithms that exploit advanced computational capabilities, in support of Air Force scientific computing interests. For the most part, this program seeks to develop innovative methods and algorithms that improve modeling and simulation capabilities. These improved capabilities, in turn, enable understanding, prediction, and control of complex physical phenomena crucial to the Air Force.

TABLE 3.3 Continued

PROGRAM	WEBSITE
Description and Funding	

Artificial Intelligence

The timely management of information, and the ability to make decisions based on that information, is of paramount importance within this program. The key issue that we are addressing is how to effectively incorporate all available information, from diverse sources and modalities, into the decision process. To understand this issue, we are sponsoring research into ways to make the best use of uncertain information; share and disseminate information; increase the accuracy, speed, and economy of the recognition and identification process; and aid the intelligence analyst.

OFFICE OF NAVAL RESEARCH (ONR) Available at: www.onr.navy.mil

Physical Sciences Division

Physical Chemistry

This program develops relationships between fundamental atomic and molecular properties and macroscopic physical states and chemical reactivity. Primary areas of interest include research relating to the chemistry and physics of sliding contacts and the development of a molecular-level understanding of lubrication.

Atomic and Molecular Physics

The research supported under this program seeks primarily to understand and exploit the interactions of light, atoms, ions, and molecules. Areas supported at present include laser cooling and trapping of both ions and neutral atoms; Bose-Einstein condensation; atom laser concepts; coherent control, e.g., of chemical reactions; femtosecond pulse shaping; quantum computing; and quantum state teleportation.

Electrochemical Science and Technology

This program develops a foundation of electrochemical sciences and technology and exploits this knowledge for energy and power systems. Areas of interest include new electrode and catalyst materials and processing, interfacial structure and dynamics, and small domains.

Solid State and Materials Chemistry

This program supports cutting-edge basic science in Solid State and Materials Chemistry (SSMC), which has a strong connection with potential Navy and DOD applications. The program is interested in revolutionary concepts and does not support incremental, evolutionary broad-based science unless it is intimately connected with a Navy problem.

Polymer Chemistry

This program supports fundamental research on the nature of macromolecules with emphasis on the solid state. The overall goal is the achievement of advanced properties that can be utilized in improving Naval systems. The average award is $100,000.

Materials Science and Technology Division

Coatings, Corrosion, and Oxidation

This task develops corrosion control approaches founded on the use of new materials, new coatings, passive films, and resistance to environment-induced cracking.

Tribology and Adhesion Science and Technology

The objective of this task is to increase the understanding of tribological mechanisms and to develop adhesion technology to enable a broader application of advanced structural and specialty materials.

Material Processing Science and Technology

This task supports research and development in materials processing and component fabrication technology to increase the flexibility and affordability of production.

continued

TABLE 3.3 Continued

PROGRAM	WEBSITE
Description and Funding	

National Science Foundation (NSF) www.nsf.gov

Mathematics and Physical Sciences Directorate

Analytical and Surface Chemistry

    Supports fundamental chemical research directed toward the characterization and analysis of all forms of matter.

Inorganic, Bioinorganic, and Organometallic Chemistry

    Supports research on synthesis, structure, and reaction mechanisms of molecules containing metals, metalloids, and non-metals encompassing the entire periodic table of the elements.

Experimental Physical Chemistry

    Supports experimental investigation of the physical properties of chemical systems. Scientific issues range from the nature and properties of individual molecules to the behavior of molecules in the aggregate.

Theoretical and Computational Chemistry

    Provides support for theoretical and computational research in areas of electronic structure, statistical mechanics, simulations and modeling, and chemical dynamics.

Division of Materials Research

Metals (MET), Ceramics (CER), and Electronic Materials (EM) Program

    This program focuses on fundamental research in the areas of metals, ceramics, and electronic materials. Projects are composed primarily of experimental activities but may incorporate some related theoretical and computational research.

Materials Research Science and Engineering Centers (MRSEC) Program

    MRSEC are supported by the NSF to undertake materials research of scope and complexity that would not be feasible under traditional funding of individual research projects.

Division of Civil and Mechanical Systems

Solid Mechanics and Materials Engineering (SMME)

    The SMME program links the expertise of analytical, computational, and experimental solid mechanics and iomechanics with materials and surface engineering to understand, characterize, analyze, design, and control the mechanical properties and performance of materials and devices.

Surface Engineering and Material Design (SEM)

    The SEM program element supports generic research on links between microstructure design and control and properties, performance, and engineering of materials and surfaces for novel applications in civil and mechanical systems and components.

Strategic Environmental Research and Development Program (SERDP)

    SERDP is the DOD's corporate environmental research and development (R&D) program, planned and executed in full partnership with the DOE and the EPA, with participation by numerous other federal and non-federal organizations. SERDP identifies, develops, and transitions environmental technologies that relate directly to defense mission accomplishment.

    SERDP is improving mission readiness through environmental research by accelerating cost-effective cleanup of contaminated defense sites; facilitating full compliance with environmental laws and regulations; enhancing training, testing, and operational readiness through prudent conservation measures; and reducing defense industrial waste streams through aggressive pollution prevention.

    Within its broad areas of interest, SERDP focuses on DOD needs in four thrust areas: Cleanup, Compliance, Conservation, and Pollution Prevention technologies.

There are also two major international organizations that provide information on D&D science and technology. The Nuclear Energy Agency (NEA) of the Organization for Economic Cooperation and Development (OECD) has established a large information exchange program, the Cooperative Agreement on Decommissioning. The International Atomic Energy Agency holds workshops and publishes safety standards, guidelines, and technical reports on up-to-date D&D technologies. Additionally, the North Atlantic Treaty Organization funds scientific research, some of which may be applicable to D&D (NATO, 2000a, 2000b), as may be some of the projects funded by the International Science and Technology Center (ISTC). ISTC projects are led by investigators in former Soviet countries that now comprise the Commonwealth of Independent States (ISTC, 2001).

The EC Research and Technological Development Framework Program

Since 1979 the EC has conducted four successive research and development (R&D) programs on the decommissioning of nuclear facilities.[4] The main objective of these programs has been to establish a scientific basis for the safe, socially acceptable, and economically affordable decommissioning of obsolete or redundant nuclear facilities. These European Union-wide programs were partially or totally funded by the EC under its five-year framework programs. The first two five-year programs focused on laboratory-scale research and development, while the latter programs included pilot-scale projects and establishment of information databases. The pilot-scale projects emphasized testing and demonstrating new technologies under representative industrial conditions to determine their feasibility, cost, and safety. Total funding since the program began has equaled about $65 million (Forsström, 1999). Table 3.4 lists the pilot projects. Within the European Union there is experience in the D&D of a variety of types of nuclear facilities, including research laboratories and reactors, reprocessing plants, and nuclear power plants.

The EC was most active in funding D&D research from 1989 through 1993. Funding has decreased as new technologies have reached maturity, and program officials feel that little additional research is needed (Colquhoun, 1999). Nevertheless, research in different topics directly connected to D&D (e.g., waste minimization or conditioning and handling of mixed waste) is still supported (EC, 1998). The committee noted that the EC approach was similar to the gate approach (see

---

[4]The European Commission (EC) is the executive body for the European Union, which includes 15 European countries.

TABLE 3.4 European Commission Pilot Dismantling Projects

Facility Name	Type	Location	Production Capacity	Production Period	Start Decommissioning Period
BR-3	Pressurized water reactor	Mol, Belgium	11 Mwe	1962-87	1989
WAGR	Advanced gas-cooled reactor	Windscale, United Kingdom	33 Mwe	1962-81	1981
KRB-A	Boiling water reactor	Gundremmingen, Germany	250 Mwe	1966-77	1982
AT-1	Reprocessing plant	La Hague, France	2 kg/d	1969-79	1981

Figure 3.1) used by OST as described in presentations to the committee (Boyd, 2000; Meservey, 2000; Brouns, 2000).

The OECD Cooperation Agreement on Decommissioning

This cooperative agreement provides an intensive exchange of information but does not financially support D&D research (NEA, 1998). This agreement began in the 1980s with ten projects from a few countries involved (NEA, 1996). Currently it includes 39 decommissioning projects in 13 countries. Membership includes the European countries, Canada, Japan, and the United States. Countries with observer status include former Eastern Bloc countries (e.g., Slovak Republic, Estonia) and Korea. Participating projects are classified as reactor and non-reactor facilities.

This agreement is the most comprehensive international information exchange about D&D technologies and progress in related areas. Information exchanges require a confidentiality agreement so that true state-of-the-art or propriety technologies can be disclosed.

Finding and Recommendation on International D&D Collaborations

There are significant opportunities for international collaborations in D&D research. In spite of this, U.S. involvement, as described to the committee, is minimal (Bedick, 2000; Menon, 2000).[5]

The committee recommends that EMSP pursue partnerships or cooperation in international research programs. These interactions should include information sharing, conferences, jointly funded research projects, and exchange of personnel at the scientific staff level.

---

[5]Present involvement includes DOE participation in committees of the OECD Cooperative Agreement, and a few DDFA projects (DOE, 2000f).

4
Research Recommendations

In this chapter the committee offers its views and recommendations on research opportunities for the Environmental Management Science Program (EMSP) to best affect important deactivation and decommissioning (D&D) problems not addressed effectively by existing technologies. Based on its discussion of future D&D challenges and their underlying causes in Chapter 2, the committee concluded that the most significant needs and opportunities lie in the characterization and decontamination steps of the D&D process, robotics and intelligent machines to enhance worker safety, and the scientific basis for determining objectives of a D&D project (facility end states). The committee has been selective in this identification to encourage the EMSP to concentrate its limited funding in a few specific areas where the committee believes research can make the most significant contributions to meeting future D&D challenges. Some technology areas, although clearly important, were excluded because in the committee's view the science and technology base already exists to address them on a relatively short time scale—less than five years.

Each recommendation is illustrated with a brief discussion of the current state of art, technology gaps, and research opportunities. Examples are included for illustration, but these should not be construed as the only opportunities that may be envisioned by the research community. Although the selection of examples was influenced to some degree by the backgrounds and expertise of the committee members, the research recommendations were arrived at by a consensus process that considered input to the committee, the needs of the end user,[1] the existence of critical knowledge gaps, the potential for future cost savings, and the possibility of achieving technology breakthroughs.

---

[1] End users are those who will use a given method or technology to accomplish a D&D task. They are usually contractor personnel at DOE sites.

Characterization

Characterization of contaminated materials, such as concrete, stainless steel, and packaged wastes, is required at nearly every stage of D&D (see Table 2.2). Labor-intensive sample collection and measurement methods expose workers to radiation and other risks and contribute to the high costs of characterization, presently estimated at 15 to 25 percent of the total D&D budget (Hart, 2000). Initially an assessment must be made to determine the types and quantities of radionuclide and chemical contamination present to ensure that adequate precautions are taken to protect workers and the environment and to assess options for eventual cleanup and disposal. Progress during decontamination must be monitored, which requires repeated characterizations during the work. Finally, for disassembly or demolition the nature and extent of residual contamination must be assessed to determine the final classification and disposal pathway of the material in question. Characterization of very low levels of residual contamination is required if a facility is to be reused.

The varied nature of Department of Energy (DOE) facilities (e.g., reactors, reprocessing canyons, laboratories, infrastructure, support) has led to a wide range of contaminant types and site-specific characterization challenges (see Chapter 2). In each case, characterization generally requires a detector tailored to the contaminant being measured and its matrix, for example, concrete, metal, liquid, or air (Janata, 1989; Webster, 1999; DOE, 2000b). In many instances, reliance is placed on characterizing and mapping sites by physically removing samples (e.g., wipes, cores), sending these to an offsite lab, and conducting chemical analysis and physical characterization. When onsite measurements are obtained, use is typically made of handheld monitors, and the data are recorded manually. Often, measurements must be repeated several times at each step of the D&D process.

As an example, over 400,000 survey measurements were made in the course of decommissioning the Fort St. Vrain commercial power reactor. Over half (221,000) were required for the final survey, which required 22 months to complete. In addition, the allowable levels of residual contamination had to be reduced by about 25 percent below the regulatory guide to account for nuclides such as Fe-55 and tritium that could not be detected with available field instrumentation (Holmes, 2000).

The committee has three recommendations for research that could lead to new or improved methods of characterizing the contaminated construction materials that are found in most obsolete DOE facilities: characterization of surfaces, characterization beneath surfaces (depth profiling), and remote mapping of contamination.

Characterization of Surfaces

The committee recommends basic research leading to ultra-sensitive devices for rapid characterization and certification of amounts of radionuclides and EPA-listed substances[2] on the surfaces of construction materials and equipment (e.g., pumps, motors).

Current Status

There are in excess of 180,000 tons of metal as well as significant quantities of machinery and equipment in surplus DOE buildings and temporary storage facilities comprising some 65 million square feet of floor area. Most surfaces are contaminated or potentially contaminated. As noted previously, these surfaces must be characterized repeatedly throughout the D&D process. The greatest opportunities for research are in characterizing surfaces bearing very low levels of radionuclide and chemical contamination. For example, before these materials can be disposed of, they must first be characterized and, if certified to be below a derived concentration guideline level (DCGL) for a specified radionuclide, may be free-released for recycling and reuse (MARSSIM, 2000). If not, they must be decontaminated and re-measured before release or, alternatively, sent to an appropriately licensed waste disposal site. Although most of the contamination is located on the surface, a significant quantity lies in microcracks and other surface defects. In addition, the presence of oils, moisture, rust, and dirt impedes the detection of low-energy alpha and beta emissions. Complicating characterization are certain radionuclides that will be essentially impossible to measure at the DCGLs in situ using current state-of-the-art instrumentation and techniques because of the types, energies, and abundance of the radiation (MARSSIM, 2000). Examples include very-low-energy pure beta emitters such as ^{3}H and ^{63}Ni and low-energy photon emitters such as ^{55}Fe and ^{125}I. In these instances, wipes or samples must be taken and analytical chemical methods used.

Opportunities

The difficulties and limitations in characterizing low levels of contamination on surfaces are important gaps in current technology. In addition faster methods that give better spatial resolution than current wipe

---

[2]EPA-listed substances are designated as hazardous by the Environmental Protection Agency (EPA). They are regulated by EPA under provisions of the Resource Conservation and Recovery Act of 1976 and its amendments.

or core methods are needed. To address these needs the committee suggests that the EMSP give special attention to research proposed for surface characterization by analytical chemistry methods, nanosensors, and biosensors.

Analytical chemistry methods

Modern methods of chemical analysis that can detect small numbers of molecules are often capable of measuring far lower concentrations of radionuclides than direct radiation measurements.[3] There are opportunities for research to extend analytical chemistry methods to develop more sensitive methods to measure small amounts of long-lived radionuclides.

Laser ablation mass spectroscopy (LA/MS) is an example of a rapid characterization technique that is suitable for solids such as concrete and requires no sample preparation (Van Hecke and Karukstis, 1998; DOE, 1998b). An intense pulsed laser is used to vaporize surface material (ablation). An inert carrier gas (argon) transfers the ablated material to an inductively coupled plasma torch, where the sample plume is disassociated into ionized atomic species. A mass spectrometer subsequently identifies the species and determines its abundance in the sample. Continued ablation of the surface provides the possibility of obtaining a three-dimensional profile of the contamination, as discussed in the next recommendation.

Sensitivity and dynamic range are such that constituent concentrations of most elements in the periodic table can be measured from parts-per-billion to tens of percent with a single analysis. The sensitivity of this analytical chemistry approach is many orders of magnitude better than can be achieved by direct radioactivity measurements for most radionuclides. Typical potential minimum detectable levels are 1×10^{-9} pCi/g for ^{238}U, 10^{-3} pCi/g for ^{239}Pu, 1 pCi/g for ^{137}Cs, and 10 pCi/g for ^{60}Co (MARSSIM, 2000, p. H-35). The technique is applicable to organic and inorganic species. Variations of laser ablation spectroscopies are attractive as well. These include laser ablation inductively coupled plasma atomic emission spectrometry (LA-ICP-AES) and laser-induced breakdown spectroscopy (LIBS). Research to adapt LA/MS and related approaches for D&D applications that require ruggedness, portability, and high sensitivity would most likely involve basic principles of energy beam-material interactions, including energy coupling, mass removal by vaporization and ablation, particle generation, gas dynamics, solid vapor entrainment, and transport processes.

---

[3] See, for example, Chapter IX, "Radiation Measurement" of Webster, 1999, and Appendix H of MARSSIM, 2000.

Research toward the development of simple, inexpensive new chemical-sensing materials is another opportunity for applying analytical chemistry methods to D&D needs. The development of simple multi-analyte chemical-sensing color test strips, currently funded by the EMSP in the D&D environmental category, is based on a self-assembled nanostructured material (Asher, 2000; Holtz and Asher, 1997) displaying photonic band-gap behavior (Joannopoulos et al., 1995). Such sensing materials might be fabricated into visual color test strips or luminescent smears to sensitively and selectively report on the concentration and identity of both chemical and radioactive contaminants, including lead, uranium, plutonium, strontium, cesium, and cobalt (Asher, 2000).

Nanosensors

Nanotechnology will have a significant impact on many aspects of science and technology,[4] and the committee believes that research in support of nanotechnology development deserves special attention by the EMSP. Laboratory-on-a-chip sensor research and micro-electromechanical systems (MEMS) are already being supported by the EMSP, but there are many knowledge gaps and opportunities for research in this new field (NSTC, 1999).

The ability to manufacture materials with switchable molecular functions will provide completely new approaches to miniaturized valves, pumps, and chemical separations and detection. Nano-composite materials also display unique mechanical, electrical, magnetic, and chemical properties. The quantum-confined behavior of thin films and nanocrystals of semiconductors has opened the possibility of designing composite materials having desirable electromagnetic properties. In principle, it is thus possible to quantum engineer improved solid-state radiation detectors. Semiconductor nanocrystals (quantum dots) in glass have recently been shown to have better thermoluminescence dosimetry (TLD) performance than current state-of-the-art TLDs (Justus et al., 1999a). It is also potentially feasible to manufacture nano-robots having a simple form of locomotion and an ability to sense and harvest targeted chemical species. These would be placed on the surface of metals to migrate into and out of microcracks as they sense and remove specific radionuclides. For example, development of a fabrication method for creating individually addressable and controllable polypyrrole-gold microactuators has been reported. This microbotic arm can

---

[4]In April 1998, Dr. Neal Lane, Assistant to the President for Science and Technology, commented, "If I were asked for an area of science and engineering that will most likely produce the breakthroughs of tomorrow, I would point to nanoscale science and engineering" (NSF, 1999, p. 1).

Chapter 4

pick up, lift, move, and place micrometer-size objects, which are about the size of a biological cell (Jager et al., 2000).

Biosensors

Biosensor technology is another relatively young field (Updike and Hicks, 1967). There has been significant growth in the development and application of living cells, bacteria, yeast, and mammalian cell cultures as sensors for chemicals (e.g., mercury, chromium, toluene, polychlorinated biphenyls (PCBs), trichloroethylene (TCE), benzene), environmental stresses (e.g., ultraviolet or oxidative damage), and advantageous and deleterious pharmacological agents (xenoestrogens). Many of these biosensors are products of genetic modifications of living organisms resulting in organic systems that produce electrochemical, bioluminescent, fluorescent, or chromogenic signals once the organisms have been exposed to a specific chemical or a broad chemical class or stress-inducing agents. Such surrogate signals of agent exposure may offer real-time online sensing capacity, while others may integrate exposure in a form of biological dosimetry. In industry, stress response genes have been linked in transcriptional fusions to create enzymatic, bioluminescent, and fluorescent biosensors and have been employed in a pattern recognition format for new chemical and environmental toxicological screening.

There has been tremendous growth in development and commercialization of a broad range of biosensor devices and applications (Kress-Rogers, 1997). Modern devices can range from fiber optic (Anderson et al., 2000) and microcantilever-linked immunoassays (Moulin et al., 2000; Thundat et al., 2000) to subcellular and cellular micro-electronic systems (Simpson et al., 1998). Analytes measurable by biosensors include a vast array of organic chemicals, biochemicals, inorganics, and metals and more recently ionizing radiation (Lee et al., 2000).

There are research opportunities to fundamentally develop and evaluate biosensor organisms to detect radionuclides and chemicals that are important for D&D applications. In addition, there are opportunities for system integration to interface the biosensor with the appropriate analytical chemistry or microelectronic platforms to provide robust detectors that can be used in D&D operations.

It might be possible for biosensors to be used to characterize occluded radioactive or chemical contamination. For example, a series of stress-inducible promoters responding to ionizing radiation could be fused with the gene for synthesis of green fluorescent protein (GFP). The biosensor organism containing the GFP fusion could be passed through decommissioned reactor lines and sampled online at an outlet to quantify net GFP as an index of exposure. Such analytical capacity would take advantage of the microscopic proportions of the sensor organism to penetrate even the narrowest orifices and at the same time provide a direct measure of biological dose exposure. It may also be possible to create coupled MEMS, bio-microelectronic, or robotic sensor devices to provide both real-time and location-specific measurement of localized contaminants.

Research to integrate microelectronics and nanotechnology (Cunningham, 1998) with elements of gene array technology (Mecklenburg et al., 2000) and cellular engineering may lead to new sensor technology. Bionanosensors are recognized as a key research area in the National Science Foundation's first call for proposals related to the National Nanotechnology Initiative.[5] Such technology could create new capacity for continuous and remote monitoring in chemically and physically complex environmental and structural systems characteristic of DOE's site D&D needs.

Profiling Construction Materials

The committee recommends basic research leading to development of real-time and minimally invasive methods to characterize radionuclides and EPA-listed substances as a function of depth in construction materials, especially concrete.

Current Status

Concrete constitutes most of the volume and weight (estimated at over 27 million tons) of DOE's surplus facilities. Because of long-term exposure, the concrete is often contaminated to a depth of several millimeters beneath its surface (DOE, 2000b), and in some cases, such as for tritium, considerably deeper. Generally during D&D, concrete floors, walls, and ceilings are first characterized and mapped by physically removing samples and sending them to a laboratory for analysis. Once the degree of contamination is determined, an effort is made either to decontaminate the affected portions or to remove near-surface material as shown on the cover of this report. The development of minimally and non-invasive real-time in situ sensing technologies to characterize the concentration of contaminants, as a function of depth within concrete, could eliminate difficulties associated with core sample collection and subsequent analysis and greatly reduce characterization costs.

---

[5]http://www.nano.gov/press.htm.

There are presently no real-time non-invasive means available to adequately determine the concentration depth profile of contaminants in concrete, hence there are good opportunities for research in this area. Gamma-ray spectroscopy has been used with limited success for selected isotopes but is not applicable for all radionuclides of relevance. X-ray fluorescence is limited to the measurement of surface contaminants.

Opportunities

There is need and opportunity for research to extend currently available or develop new surface analysis methods to provide a full three-dimensional picture of contaminants as a function of depth in construction materials. Research to extend laser ablation mass spectroscopy (LA-MS) along the lines described under characterization of surfaces is one example (Van Hecke and Karukstis, 1998). Research toward MEMS or bioorganisms that could tunnel into construction materials and provide contaminant profiles may offer significant potential. Contaminant profiling is an important need for D&D for which there appears to be little current applicable research.

Remote Mapping

The committee recommends basic research leading to the development of methods for remotely mapping radionuclides and EPA-listed substances.

Current Status

Remote-sensing systems can provide both economic and safety benefits by distancing the worker from hazardous work areas. The EM Office of Science and Technology (OST) and its deactivation and decommissioning focus area (DDFA) have recognized this advantage. Remote mapping of activity levels using gamma cameras is a recent innovation that has proven useful in D&D operations. These instruments display the relative strength and location of gamma radiation as a two-dimensional image superimposed on the corresponding visual image (DOE, 1998c, 1998d). However, there are still significant gaps in this technology because it is not able to survey areas with low levels of radiation (e.g., for release surveys). Nor are the current systems able to identify specific radioactive isotopes. Isotopes that primarily emit beta particles (e.g., tritium, technetium-99) or alpha particles (e.g., plutonium-239) are generally much harder to detect and quantify at low levels than are gamma-emitting isotopes (DOE, 1999a).

Limitations of the current technology stem from the use of scintillator materials in the detector head that lack the ability to distinguish between gamma rays of different energies and, consequently, specific radionuclides. Currently available solid-state detectors, on the other hand, can distinguish energies but suffer other drawbacks, including poor sensitivity and the need for cryogenic cooling.

Opportunities

The committee believes that current technology for remote mapping can be significantly advanced by research in new detector materials, fiber optics, and fiber optic-based dosimeters.

Detector materials

There are research opportunities to discover or synthesize new detector materials to overcome current limitations (Squillante and Shah, 1995; Entine and Squillante, 1999). In general, a good material must be readily available in high-quality crystal form and be compatible with electronic device fabrication procedures. In addition, the material should strike an adequate balance between physical properties such as charge carrier mobility, carrier lifetime, resistivity, resistance to radiation damage, and stopping power, all of which contribute to detection sensitivity. Recent advances are promising. Room temperature detection has been reported using silicon (Si), cadmium telluride (CdTe), and cadmium zinc telluride (CZT). Other new materials such as GaAs, TlBr, and PbI have attracted attention.

Device geometry may also be altered to improve performance. For example, relatively large output pulses are produced with an improved signal-to-noise ratio when a p-n junction in a silicon wafer is used in an avalanche-photodiode geometry. Silicon drift detectors with concentric-ring geometry have inherently low capacitance, also improving performance significantly. In principle, enhanced radiation absorption and sensitivity may be obtained through the use of quantum-confined nanostructured detection materials containing quantum wells or dots. These composites are made from existing semiconductor materials, like GaAs, that are already extensively used in electronics but have properties that differ dramatically from their bulk crystalline counterparts. Their characteristics, especially their electronic and magnetic properties, are quantum dominated. Such devices are already being quantum engineered for uses in optics, electronics, and magnetics. The committee believes that research opportunities exist in the development of new solid-state detector materials and geometries with significantly improved radiation-sensing properties, especially nanostructured materials with desirable quantum-engineered behavior.

Fiber optic chemical sensors

Fiber optic sensors are another example of remote sensor technology in need of further research. Fiber optic chemical sensing has been explored (Lieberman, 1996) and probes to measure various chemical species in varying environments (e.g., air, water, soil, in vivo) have been developed for environmental sensing and medical applications. Most often a coating that has been formulated to selectively absorb the chemical species of interest is applied to the outside cladding of the fiber. As the coating selectively absorbs the species of interest, the light transmission through the fiber is reduced. Fiber optic chemical sensors are already employed at former nuclear weapons sites primarily for subsurface soil characterization and water analysis. For example, fiber optic sensors have been used at the Hanford Site to monitor carbon tetrachloride in water at parts-per-million levels and at the Savannah River Site to measure trichloroethylene at parts-per-billion levels.[6] Further research is needed to develop coatings that are specific to other species of D&D interest.

Fiber optic radiation sensors

Remote sensing of radiation using optical fibers as the detector material and a conduit for transmitting accumulated dose information is a recent development (Henschel et al., 1994; Huston et al., 1996; Borgermans et al., 2000). Opportunities exist for both improved performance and such novel features as optical interrogation (Miller et al., 2000). Fiber optic radiation sensors would find considerable use in surveillance and monitoring, for example, in fiber optic pipe surveys, in cone penetrometers for assessing the migration of radioactive species in the soil around buildings, and in monitoring liquid tank leaks. Early versions of fiber optic sensors relied on damage and the resultant decrease in transmission (darkening) with accumulated dose that occurs in glasses exposed to intense radiation (Henschel et al., 1994; Evans et al., 1978). The composition of glasses can be modified using a variety of dopants to significantly alter the radiation sensitivity from fractions of a Gy to 10^5 Gy (Beuker and Haesing, 1994; Tomashuk et al., 1999). In practice, rapid fading of the radiation-induced darkening complicates the determination of the absorbed dose.

Fiber optic luminescent sensors have recently been developed based on optically transparent electron-trapping glass materials containing low concentrations of metal ions such as copper, cerium, or europium and clusters of nanocrystals such as ZnS or Cu_2O (Justus et al., 1999a, 1999b). Radiation exposure results in trapping of electrons at defect

---

[6]http://www.srs.gov/general/srenviro/erd/technology/a06.html.

centers within the glass matrix. Entrapment can persist for years at room temperature, so this approach has a low fade rate. The trapped electrons can be stimulated to recombine with the ionized centers by heating (thermal luminescence, TL) or by illumination with near-infrared light (optically stimulated luminescence, OSL). The recombination process results in a luminescence signal that is proportional to the absorbed radiation dose. Moreover, after TL or OSL, the fiber is refreshed and can be reused, a requirement for permanent installation in a remote monitoring system.

These materials have demonstrated greater sensitivity than TLD materials currently in use and are remarkably linear with accumulated dose (over seven orders of magnitude) while exhibiting low fade rate. The glass can be drawn into long lengths of fiber and fused directly to commercial optical fibers (Huston et al., 1996; Miller et al., 2000). A fiber sensor based on OSL has been fabricated that provides a sensitivity of 10^{-4} Gy per meter (Miller et al., 2000). The doped glass materials can also be used in a variety of non-fiber forms such as chips, rods, or powders for other radiation monitoring applications (e.g., personal badge dosimeters). Improvements in both materials and analysis methods are required to develop reliable, robust fiber optic dosimeters for DOE radiation monitoring activities.

Decontamination

Like characterization, decontamination of equipment and facilities is necessary at most stages of the D&D process (see Table 2.2). Initially, radiation and contamination levels may have to be reduced to allow workers access or to limit their exposure to radiation and other hazards. Decontamination may be required before dismantling or demolition work to prevent the spread of radioactive or toxic materials, which can have adverse offsite as well as onsite consequences. Decontamination procedures are intended to result in a small volume of the most hazardous waste, and much larger volumes of waste that has low or no hazard, thus reducing the cost and long-term risk of disposal. Some decontaminated equipment or facilities might be recycled or reused. The end state of any decontamination activity must be consistent with both site-specific and overall DOE cleanup objectives.

For a D&D project, considerations of cost, schedule, and worker safety will lead to an optimization of how much time and effort should be put into decontamination efforts versus simple waste conditioning and disposal. For example, at present, bulk waste materials that are decontaminated to USNRC Class C or below can be disposed of at relatively low cost in disposal facilities on DOE sites, whereas materials greater than Class C will go to a special repository. There are significant uncertainties in future waste disposal costs, however, which provide an incentive for research that may make decontamination faster and more effective.

For large-scale applications, almost all current decontamination methods are time consuming, involve risks to workers, produce significant volumes of secondary waste, and often leave residual contamination, especially actinide contamination. They usually require direct, hands-on work such as the concrete spalling work shown on the front cover of this report. Other available methods include wiping the surface with cleaners (e.g., detergents, acids, complexants), washing with high-pressure water, immersing objects in various cleaners, and electro-polishing (which removes a thin layer of the surface of metals). Two other methods in use are blasting with solid CO_2 or sodium carbonate.

With current technologies D&D contractors usually choose to send large amounts of contaminated materials (e.g., concrete and steel) to licensed disposal facilities rather than attempt to decontaminate them for possible reuse. However, even for concrete, a relatively cheap raw material, recycling can be economical (Parker et al., 1998). Presentations to the committee indicated a need to improve current technologies for removal of radionuclides and EPA-listed organics and metals from equipment and building structures and metal, concrete, and wood debris. In many instances, paints, sealers, and varnishes create a laminate problem, with aged materials being harder to decontaminate than more recent deposition. Deep penetration of contaminants into porous structural material, such as concrete, also makes decontamination difficult.

Fundamental Interactions and Modeling

The committee recommends basic research toward fundamental understanding of the chemical and physical interactions of important contaminants with the primary materials of interest in D&D projects, including concrete, stainless steel, paints, and strippable coatings. Results should be used to develop first-principle models that describe the interactions and can be used to investigate improved approaches to decontamination.

Current Status

While there exists a good deal of chemical data on the contaminants themselves (Delany and Lundeen, 1990) and on their transport in the environment (van der Lee et al., 1997), there is little information of direct relevance to D&D problems. Both radioactive and toxic contaminants are known to exist in a variety of chemical forms (e.g., in different valence states, complexes, or as colloids) that exhibit very different behaviors.

Modeling of radionuclide and chemical contaminant behavior that is relevant to D&D problems is almost nonexistent. Conversely, there has been extensive modeling work in other DOE problem areas, such as subsurface contamination and high-level waste disposal (NRC, 2000a, 2001a). Available models are not adequate for developing improved decontamination processes. For example, surface oxides are known to sorb metal ions. The sorption has been described by a wide variety of models. Most of these models are based on measurements taken under a specified set of conditions (e.g., sorption isotherms) rather than on fundamental parameters. They are not generally applicable to the variety of conditions encountered in decontamination activities. In most studies neither the surface nor the metal ion is explicitly treated. Often, the role of the chemical form of the contaminants (speciation) is neglected.

Present decontamination approaches are usually based on experience or trial and error rather than quantitative prediction of how the contaminants are bound to construction materials and how chemical or physical methods can best remove them. Currently trial and error is often the only game in town because the history of how a facility was contaminated is unknown (see Chapter 2) and available methods to characterize the contamination are not adequate.

Actinide-contaminated materials are a major problem at many DOE sites. Contaminated materials include glove boxes, shielded cell liners, concrete, lead bricks, lead glass, and plastics. Radioactively contaminated lead, which is also chemically toxic, is a particular challenge. The DOE has a large inventory of contaminated lead due to its use as shielding material. The ability to efficiently remove actinides from the surface of construction materials will allow recycling or cheaper disposal of the material.

Opportunities

Scientific understanding of the interactions among contaminants and construction materials is fundamental to developing more effective D&D technologies. Such information includes how contaminants bind to steel and concrete surfaces; how they penetrate into these materials; their migration into pores, fissures, and welds; and time-dependent aging effects (Dzombak and Morel, 1990).

Decontamination studies should focus on a few fundamental parameters and interactions that can provide useful data for developing new decontamination methods. Investigations on radionuclides particular to the DOE, such as actinides, should be stressed. A variety of conditions (pH, temperature, ionic strength) should be examined. The interactions should be kinetically and thermodynamically described to facilitate applying the data to a variety of decontamination conditions and to ease the incorporation of data into first-principle models.

Modeling from first principles provides an opportunity to integrate the results of fundamental research in both chemical interactions and biological processes that are relevant to D&D problems. Such models can be the first step in bringing new knowledge to bear on improving decontamination approaches and processes. Properly integrating chemical and radionuclide speciation into D&D models is likely to be especially informative—adding new knowledge in general—since the most important species will likely be different from those in high-level waste or in subsurface contamination because of their different chemical environments. Beyond their use in decontamination the models can help provide a more general scientific basis for predicting behavior of contaminants in construction materials as a scientific underpinning of facility end states, which are described in the final recommendation in this chapter.

Biological Processes

The committee recommends basic research on biotechnological means to remove contaminants from surfaces and from within porous materials found in surplus DOE facilities.

Current status

The capacity of microbiological processes to destroy, transform, mobilize, and sequester toxins, pollutants, and contaminants is well established (Young and Cerniglia, 1995). Microbially produced surfactants and bioemulsifiers have been examined as replacements for organic chemicals in surface cleaning, stripping, and remediation. Microbial processes have been investigated for coal and tar liquefaction. A variety of microbial process activities are documented in the degradation, transformation, and immobilization of organic and inorganic chemicals and radionuclides. Bioleaching is a well-developed technology that has long been practiced for metal ore processing and has been developed for valuable metals recovery. Recent work at the Idaho National Engineering and Environmental Laboratory has employed bacteria for surface decontamination studies (Rogers et al., 1997). DOE has recognized the potential value of biotechnical remediation methods and supports research through its Natural and Accelerated Bioremediation Research (NABIR) and Subsurface Science programs as well as the EMSP (DOE, 1995).

Opportunities

The committee believes that the rapidly growing field of biotechnology affords many opportunities for research that is relevant to DOE's facility decontamination needs. In addition, a fundamental understanding of the biological processes would help to ensure that waste by-products from the decontamination could be minimized, safely treated, and stabilized. Other research opportunities lie in determining effective ways to apply biological agents. For example, can high pressure be used to infuse bacteria or required nutrients into and through a contaminated concrete matrix? Biological treatment of in-place structural materials may differ from treating debris rubble. Biotechnical research areas that the committee recommends for emphasis include bioleaching, biosurfactants, and biocatalysts. A few examples of specific research opportunities are also given.

Bioleaching

Research to extend well-known technology in mineral ore leaching and metal recovery (Torma, 1988; Ehrlich and Brierley, 1990) can provide biochemical capacities for removal of metals and radionuclides from construction materials. Applications for contaminated concrete could be profound if volumetric reductions were achieved.

Bioleaching generally uses a Thiobacillus-driven production of H_2SO_4 and a Fe^{++} to Fe^{+++} couple to create a leaching solution. Numerous metals, including uranium, can be recovered commercially. Metal-rich ore differs significantly from concrete because the pH of the ore leaching solution is relatively easily maintained at pH <2. Concrete, being highly alkaline and buffered, poses difficulty for establishment of sulfur-oxidation-driven bioleaching, but alkali-tolerant strains of Thiobacillus have demonstrated potential. Further exploitation of such organisms and nitrification-driven nitric acid concrete leaching is warranted.

Biosurfactants

There are opportunities for research that could lead to the use of microorganisms in aqueous or surfactant mixtures (with or without solvents or other surface-tension-lowering agents) for surface treatment and contaminant removal. Analogous technology has been developed for problems of soil bioremediation. In application for structural materials and debris it may be anticipated that previous problems (such as loss of solvents and surfactants due to mass loadings on soils or poor recovery and recirculation) can be avoided due to the different nature of the contamination matrix. In general, the area of biosurfactants and emulsifiers is under-investigated and could have broad implications not only for the EMSP but also for defense and industry (Sullivan, 1998; Banat et al., 2000).

Biocatalysis

Uranium can be microbially mobilized to the hexavalent form aerobically and can be reduced and precipitated (even from tridentate organic complexes) as tetravalent U by sulfate-reducing bacterial process technology. Other radionuclides and metals can likewise be precipitated as sulfides. Similarly, a variety of hazardous contaminants (e.g., PCBs, TCE, and perchloroethylene) can be independently or cooperatively degraded by joint aerobic and anaerobic bacterial processes with or without synergistic action of physical-chemical treatment, such as photocatalytic oxidation, or Fe^{++}-driven reduction. Process technology has been created that couples such treatments with surfactant or solvent flushing and washing. While originally developed for hazardous waste soil remediation, technologically these processes may be well suited, perhaps better suited, for more defined applications in decontamination.

Examples

Research in the following areas, in the committee's opinion, is most likely to lead to relevant new knowledge and technologies for DOE facility decontamination problems.

- selection, isolation, or biological engineering of acid, alkali, and solvent-optimized bacteria for focused application in decontamination;
- cell-free enzyme development for decontamination, including gel-polymer immobilization matrices;
- improved kinetics for aerobic and anaerobic biocatalytic decontamination;
- improved resistance to chemical matrix toxicity; and
- microbial (bacterial and fungal) production of extracellular chemicals and enzymes for application in surface cleaning and delamination.

Robotics and Intelligent Machines

DOE has recognized the potential for robotics and intelligent machines (RIM) to meet some of its greatest challenges, including

- lowering the cost of operations in the face of budgetary pressures,
- meeting the rising regulatory standards and continuing to improve the health and safety of its workers, and
- carrying out necessary operations that are too hazardous for humans to perform.

Because RIM crosscuts almost all of DOE's programs, DOE has laid out long-range plans for developing this technology in a Robotics and Intelligent Machines Roadmap (Sandia, 1998). Technology roadmaps are planning documents that DOE uses to call attention to future needs for development in technology, provide a structure for organizing technology forecasts and programs in order to avoid gaps or overlaps, and communicate needs and opportunities throughout the R&D community. For EM activities, the roadmap lays out ambitious goals for RIM, which include

- increasing productivity by 300 percent,
- reducing personnel exposure by 90 percent, and
- reducing secondary waste by 75 percent.

The DDFA estimates that 30 percent of its needs include RIM requirements (Haley, 2000). Facility D&D presents workplace hazards that are unique among EM's challenges. Most D&D baseline technologies require that workers routinely enter areas with radiation and many other industrial safety hazards and perform hands-on work with powerful and heavy equipment, including cutting devices that can instantly penetrate protective clothing. This routine work includes sampling (for characterization), decontaminating, and eventually dismantling the barriers that were originally constructed to protect workers from radioactivity and toxic chemicals.

As discussed in Chapter 2, DOE facilities are massive, they are often crowded with complex, heavy equipment, and in many cases details of how the equipment was designed and operated have been lost. The reality of nuclear facility D&D is that the physical tasks are unstructured (not repetititve) and involve a wide variety of highly contaminated components (e.g., piping, valves, wiring, tanks). When performed directly by human workers this work represents a significant safety risk and a high cost in resources and time (see Appendix D). The first goal of remote systems technology is to remove the workers from harm's way, which dramatically improves safety. The second goal is to increase productivity and reduce costs and project schedules—all of which would make the D&D enterprise more manageable. Considering the present timeline for this D&D work, there is at least a decade available during which an accelerated science-based development program could be pursued to revolutionize the technology to meet the goals of the RIM roadmap.

Chapter 4

Intelligent and Adaptable Remote Systems

The committee recommends basic research toward creating intelligent remote systems that can adapt to a variety of tasks and be readily assembled from standardized modules, with special emphasis on actuators, universal operational software, and virtual presence.

Current Status

DOE's RIM roadmap lays out needs and plans to greatly advance current technologies, but the envisioned leaps in technology are not likely to occur without new knowledge. Even though industrial robot systems have reached a level of maturity providing high availability at low cost, they are not well suited to the complex and unpredictable demands of the D&D task environment. D&D has special needs primarily because it is unstructured, requiring continuous kinesthetic input and supervisory oversight, and also because the loads and motion ranges are very large and impact loading can be very severe.

Current technology available for D&D is best represented by the Chicago Pile 5 (CP-5) large-scale demonstration of a dual arm robotic system (see Table 3.2). This system used an older hydraulic control system with nominal software. Simple tasks were performed, maintenance and downtime were extensive, the rate that the work was accomplished was low, and resulting confidence in the technology was low (NRC, 1998a). The system did perform the overall dismantlement function, but it also raised the question of efficiency (cost and time). Most efforts to improve the technology have resulted in one-of-a-kind demonstrators (not based on a full architecture with components enhanced by a continuous science effort), which are hard to certify as to their real performance.

Opportunities

Research opportunities exist primarily in making RIM more humanlike in their abilities to adapt to a variety of tasks, both physically and intellectually. In this regard, the committee believes that the best opportunities for research relevant to D&D tasks involve actuators, universal criteria-based software, and virtual presence.

Actuators

The actuator is the power (muscle) of remote systems and as such it is the key to performance, reliability, and cost. Except for better materials and improved control electronics, most actuator technology has not changed for several decades. Today's actuators typically use only one sensor (for position) so that virtually no real-time data (e.g., force and velocity) are available to make them "intelligent." More complete sensory input coupled with decision-making software (see below) can produce intelligent actuators that are able to adapt to a variety of tasks. Achieving a relatively inexpensive modular design to allow plug-and-play deployment of these devices would be especially useful in D&D projects, because equipment that fails or becomes contaminated is usually discarded. Research to answer the question of granularity (What is the minimum number of required standard modules?) to enable the assembly on demand of the maximum number of remote systems would make the overall system substantially more cost effective in deployment and maintenance.

Universal criteria-based software

Criteria-based decision making is the essence of intelligence in robotic systems. Today's control of robotic devices is derived from techniques developed during World War II in which control is linear (based only on the difference between two measured parameters). A robot capable of mimicking human adaptability, however, would require a non-linear control system in terms of many highly coupled parameters corresponding to the physical features that accurately represent performance of the task. The criteria-based software could be universal in the same sense that operating systems on microcomputers are universal—one system supports many different applications (Miller and Lennox, 1991; Sturzenbecker, 1991; Stewart et al., 1992).

Virtual presence

In the initial planning and characterization phases of D&D work, workers often must enter an area of high radiation and contamination that is also congested with left-in-place equipment and materials for which removal inevitably involves physical stress (fatigue) and the potential for personal injury (Fournier et al., 1998). Virtual reality systems could allow workers to perform essential survey and decision-making functions from a remote location (Lloyd et al., 1999), thus enhancing their safety and productivity. Advances in the state of the art as, for example, in deep sea exploration could improve overall system performance by providing force feedback, remote vision, collision avoidance, and radiation-resistant sensor technology.

Summary

There are research opportunities to develop basic scientific understanding to support the new actuator, software, and virtual presence technologies that the committee believes are the keys to achieving DOE's objectives for RIM. To effectively organize this research the committee believes there is need and opportunity to establish a small number of technology demonstrations at the DOE laboratories to convincingly document measured performance of the present state of the art. This would be technology that exists or could be aggregated in the next two to three years (not one-of-a-kind systems or extensions of industrial robot technology). These demonstrations would not only show where investment in research is needed but would also provide a baseline to measure progress.

The RIM roadmap contains a list of applications and a tabulation of the desired technology (Sandia, 1998). Some of the most significant science opportunities to achieve these technology goals are listed below. A more complete discussion is given in Appendix D.

- Sensors for site characterization and acquisition of performance data are essential to support decision making either through software or by visualization and human judgment.
- A special need is the kinesthetic interface to human operators to enhance their motor skills and input commands to the remote system.
- Mobile platforms that in themselves are modular and highly dexterous must be further developed to gain access to the work environment and to transport size-reduced facility components.
- Quick-change end-effector tools based on a science of tools (design, modeling, and operation) are needed to perform the in-contact physical tasks for D&D.
- Dexterous high-load robot manipulators capable of tool management, size reduction, parts transport, and parts packaging under human supervision will be especially important for D&D tasks.
- Intelligent and standardized actuator modules to build all D&D remote systems on demand from a minimum number of high-performance and low-cost modules just as we now build computers on demand.
- A universal operating software for any intelligent machine used for D&D. Similar to the operating systems in today's microcomputers, this software would support operation of mobile platforms, gantries, small automation subsystems, and dexterous manipulators, all under the centralized supervision of operators in a remote position.
- Electronics will be pervasive in a modern remote systems technology. Hence, their hardening against radiation, temperature, shock, and particulates is necessary.

End States

The purpose of facility D&D is to leave a building in an agreed-upon condition, referred to as the end state. As discussed in Chapter 2, end states are established though input from many involved people and organizations through considerable negotiation. End states for facility D&D are usually a part of the goals for overall cleanup of a site. These goals can be general, such as to protect the Columbia River at Hanford, or more specific, such as to re-industrialize parts of the Oak Ridge Site. At Rocky Flats most buildings will be demolished and the debris removed from the site. After completing its site visits and in the continuing deliberations after its interim report, the committee agreed that there is need for a better scientific basis for evaluating the acceptability of alternative D&D end states.

Fate and Behavior of Contaminants in Construction Materials

Research should be directed toward understanding the fate and behavior of treated and untreated contaminated material by determining the fundamental chemical species of the contaminants in the host material and how the species behave. The effect of time and changing ambient conditions should be considered in these investigations. Further research should be directed at incorporating these results into risk assessments to evaluate and compare long-term safety that can be provided by different end-state options.

Current Status

Considerable effort has been directed at scientific understanding of the fate and behavior of radioactive wastes in the environment (NRC, 2000a) and, to a lesser extent, of hazardous wastes (NRC, 1999b). There is very little scientific understanding of the fate and behavior of radionuclides or EPA-listed chemicals in construction materials. Understanding how contaminants may move through or out of concrete and metal is important for providing scientific input toward establishing the end state. From a research perspective, developing this scientific knowledge is clearly linked to the committee's recommendations in the areas of characterization and decontamination.

Some of DOE's largest facilities slated for D&D have potential future uses. Evaluating the different uses requires an understanding of the long-term behavior of the contaminated material. The Hanford Site provides four examples: (1) After the end of operations for the N reactor, some buildings were returned to service for spent fuel storage (McDaniel, 2000); (2) the B reactor is being preserved as a museum; (3) building entombment for low level waste disposal is under consideration for former chemical separation plants (MacFarlan, 2000); and (4) there are plans to re-industrialize parts of the site (the 300 area).

Opportunities

Research is needed on understanding the physical and chemical forms (speciation) of contaminants in building construction materials. Buildings may be in storage for decades before D&D and potentially in use for decades thereafter. Understanding the speciation and behavior of the contaminants, how the speciation evolves with time, and the impact of decontamination activities on the chemical speciation is critical for developing a scientific basis for determining end states.

Decontamination often uses chemical or biological processes that can impact the behavior and performance of construction materials as well as the contaminants themselves. The use of chemicals or bacteria for decontamination can dramatically affect the local environment by changing pH or inducing chemical reduction or oxidation reactions through respiratory activity. This activity, coupled with physical changes due to material cutting, melting, or polishing from decontamination efforts, can impact the behavior of the contamination. For example, acids dissolve concrete; less dramatic reductions in pH can also have profound effects, but details of these changes over time and how they may affect the eventual release of contaminants are not well understood. Even if there are no decontamination activities the host material will change with time.

The behavior of hazardous airborne species (Egge, 2000) presents another opportunity for research. While water is the primary mover of contaminants in the ground (subsurface contamination), airborne pathways may be especially important in establishing a scientific basis for facility end states. Performance assessment modeling using fundamental fate and transport data could be developed as an important decision-making tool for establishing facility end states.

5

Programmatic Recommendations

In two presentations to the committee, Mark Gilbertson, Director of the EM Office of Basic and Applied Research, asked the committee to suggest a vision for the Environmental Management Science Program (EMSP) that would help the program to best fulfill its congressional charter (Gilbertson, 2000a, 2000b). In addition, the committee's statement of task provided that "[t]he committee may also consider and make recommendations, as appropriate, on the processes by which (1) future research needs can be identified, and (2) successful research results can be applied to DOE's D&D problems." In the course of its site visits, fact finding, and subsequent deliberations, the committee developed ideas that it felt would be helpful to the EMSP in the above areas. In addition the committee believes that the EMSP can strengthen itself by emphasizing its role in developing new scientists and engineers and the commercial value of the research that it funds. The committee's programmatic recommendations for the EMSP are described in this chapter.

A Vision Statement for the EMSP

The EMSP was created by the 104th Congress to "stimulate basic research, development, and demonstration efforts to seek new and innovative cleanup methods to replace current conventional approaches, which are often costly and ineffective."[1] Accordingly, EMSP research is needs driven or mission directed (NRC, 1997).[2] The EMSP has the

---

[1] Conference report accompanying the Energy and Water Appropriations Bill (Public Law 104-46, 1995).
[2] This committee as well as other review committees (NRC, 1997, 2000a; EMAB, 2001) have interpreted EMSP's charter to include basic and applied

opportunity to bridge the gap between fundamental research and needs-driven applied technology development. Accordingly the committee believes that the vision for the EMSP should extend from fundamental science to applications of science and technology that create major benefits for the EM program. The proposed EMSP vision statement is as follows:

Provide scientific knowledge to allow dramatic improvements in worker safety, cost, and schedule for meeting the national need to clean up DOE sites while protecting public health and the environment. In doing this, the EMSP will be recognized as a key partner by the focus areas and DOE sites, will be supported by Congress and stakeholders, and will be preparing and developing qualified scientists for future DOE program needs.

To help achieve this vision the committee believes that the EMSP and the rest of the EM community must develop and pursue aggressive, shared goals for improvements in worker safety, cost, and schedule. The ambitious goals for dramatic improvements (factors of 3 to 10) set forth in the Robotics and Intelligent Machines Roadmap (Sandia, 1998) are appropriate examples for the EMSP.[3] The committee believes that establishing and pursuing such shared goals will significantly enhance the EMSP's efforts to identify needs and apply results, as discussed in the next two sections.

Improving Science Needs Identification

In July 1996 DOE issued a "National Needs Assessment" document that attempted to identify the complex-wide technology needs for DOE's environmental management program. For the D&D area, the document identified 102 technology needs grouped into 31 categories. The document was compiled from inputs received from DOE field offices and represented a bottoms-up list of specific needs. There was no management top-down review, evaluation, and prioritization of these needs (NRC, 1998a). In 1999 the NRC report *Decision Making in the U.S. Department of Energy's Environmental Management Office of*

---

research, gates 1 and 2 in the Office of Science and Technology's (OST) technology development strategy (see Figure 3.1). OST's focus areas are responsible for developing, demonstrating, and deploying new technologies.

[3]These goals are not the result of the committee's analysis, nor intended to be required of contractors, but are indicative of goals that could be achievable and will help identify true breakthrough areas for research.

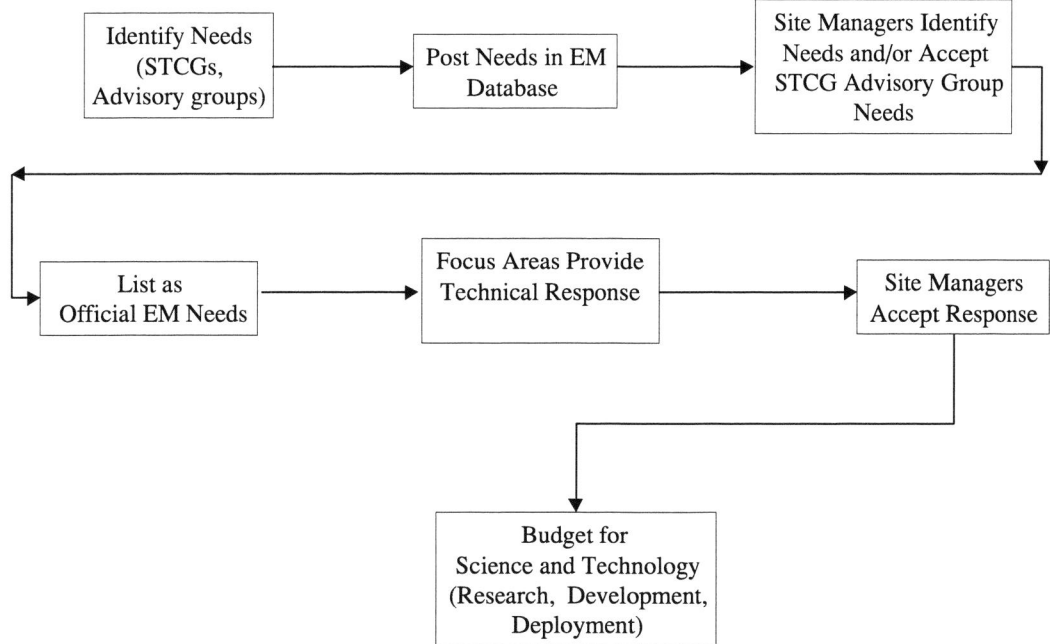

FIGURE 5.1 The OST's process for identifying and responding to EM site needs for science and technology. Source: DOE.

Science and Technology found that there were no general guidelines for setting criteria for the selection and prioritization of technology development needs (NRC, 1999e).

Since that time EM's Office of Science and Technology (OST) has significantly improved its needs identification process (see Figure 5.1), and it maintains a publicly accessible needs database.[4] Nevertheless, the committee found that there is no complete, comprehensive, and coordinated definition of D&D science needs for the DOE complex. The committee saw a variety of needs lists, particularly those from the site technology coordinating groups (STCGs), but most lists were too narrow, short term, or site specific to help determine where basic research could be helpful. Presentations made to the committee indicated that most attention and funding are aimed at short-term, site-specific D&D problems. While relevance to site-specific problems is important, too narrow a focus may preclude funding of novel outside-the-box research with potentially greater impact on D&D.

The sources of information for needs input into the EMSP, including advisory committees such as this one, are shown in Figure 3.3. Both the OST and this committee recognize that an advisory committee's findings are necessarily based on snapshot fact-finding visits to the sites. The committee believes that comprehensive and sustained needs identi-

---

[4] http://emsp.em.doe.gov/needsdatabase.asp.

fication must come from those who live with site problems on a daily basis—the contractors, field offices, and DOE headquarters personnel.

The committee has three areas of recommendations, which it develops in the following sections, to help OST and the EMSP to improve the identification and definition of D&D technology needs:

- Engage and integrate EMSP researchers and site contractor personnel.
- Develop a better framework and prioritization of technology needs.
- Provide an in-depth parametric analysis of a few key long-term D&D challenges.

These are not one-time actions, but will require follow-up as D&D work progresses at DOE sites, lessons are learned, more needs are identified, and regulations evolve. Management tools that can help ensure appropriate follow-up include OST's Multiyear Program Plan, the needs database, and especially independent review (DOE, 2001; EMAB, 2001).

Engaging scientists and contractors

Today it is recognized that even routine scientific progress usually requires the efforts of a team of scientists, engineers, and technicians having a range of skills and education. A previous NRC committee writing about the EMSP noted that "the advancement of scientific knowledge is a cumulative effort involving many scientists over long periods of time (NRC, 2000a, p. 127). Collaborations are especially important when research is expected to lead to a new deployable technology. For the EMSP to succeed in the D&D area, where site contractors are reluctant to adopt new technologies, collaborations not only among scientists—the problem solvers—but also with contractors—the problem holders—will be necessary.

In the D&D area the committee believes that the EMSP has not yet significantly engaged the scientific community. Since 1996 the EMSP has funded a total of 22 D&D research projects (see Table 3.1) covering six separate topics: inorganic chemistry, biogeochemistry, analytical chemistry and instrumentation, engineering science, separations chemistry, and materials science—thus on the average less than four projects in each of these important areas. For the past two years there has been no new funding for D&D science programs (Gilberston, 2000b). Only a fraction of the renewal applications from three-year grants in FYs 1996 and 1997 are expected to receive awards (Gilberston, 2000b). To interest scientists in helping to identify D&D research needs and to engage them on a sustained basis, stable and adequate funding will be necessary. The committee did not deal with research funding issues, but it noted that effectively engaging the scientific community would cost more than is currently being invested. Other committees are providing funding recommendations to the EMSP (EMAB, 2001; NRC, 2001b).

The committee recommends that EMSP should engage the scientific community more effectively in identifying and participating in promising areas of research. A broad sustained effort by the scientific community to understand and engage in the D&D challenges would allow them to define the scientific information needed and propose relevant topics for research. This would also attract young scientists and engineers to the D&D field at a time when the availability of scientific personal is declining.

DOE contractors currently engaged in D&D operations at the various sites are a potential source of information about difficult-to-address, long-term science-based problems. Presentations by contractor representatives at the sites visited by the committee emphasized that the methods proposed to complete the operations under their responsibility were baseline (tried and tested) technology. Innovation or experimentation with new methods, with its concurrent risks of failure, was considered outside the scope of their contract and not cost effective. EM is missing the opportunity to engage technical personnel at the front line of the cleanup effort in identifying longer-term needs and in providing guidance and assistance in developing fundamentally better ways to do the job.

The committee recommends that EM should encourage the contractors and DOE site management to take a broader, long-term perspective of D&D needs for work to be performed in ten years or more, so that technology solutions can be developed that provide greatly improved D&D operational capabilities.

The committee has two suggestions for better identifying D&D science needs by integrating research and contractor personnel: increase their interaction in the pre-proposal phase and include them in research teams. There is an opportunity for contractor personnel with firsthand knowledge of D&D problems to participate in the EMSP's biannual workshops. Their perspectives in helping to identify problems could improve the workshop discussions and subsequently the quality of research proposals that are submitted to the EMSP. Site visits by proposing teams could be encouraged in the pre-proposal stage as another means to improve proposal quality. One of the outcomes of a dialogue between contractors and researchers might be recognition of new and innovative approaches to research issues that would otherwise not be identified. It might also be useful to involve the contractors in the EMSP's pre-proposal screening and feedback process (see Figure 3.3).

To further increase involvement of the contractors, it should be possible for appropriate contractor staff to be included with the research team in the EMSP proposal, perhaps with matching in-kind contributions from the contractor. Continuing involvement of contractor personnel with the research team throughout the research program should provide positive benefits to both parties.

Framing and Prioritizing Needs

Although the OST has made substantial improvements in D&D technology needs identification, after hearing presentations from site technology coordinating groups during its fact finding, the committee feels the process is not being fully utilized and that there may be ways to further strengthen it, as follows:

Step 1. Establish a *framework* for defining problem areas requiring new science and technology. This framework can be the phases of D&D shown in Table 2.2, systems engineering functions for D&D, or other categorization of tasks. To be useful in needs identification the framework must be broadly applicable to all DOE D&D work. Then, criteria or attributes must be defined to allow description and quantification of the problems and the corresponding science and technology needs. Such criteria or attributes could include the following:

- estimate for the technology's complex-wide usage expressed in employee hours, surface area, material volume, and/or people affected;
- improvements in protection for the worker, public, and the environment;
- total cost savings;
- return on Investment;
- importance of problem solution to accomplishing the field office's site mission; and
- other attributes specific to the problem area.

Each criterion can be assigned an appropriate weighing factor, if needed. Cost effectiveness should be evaluated on a DOE-wide basis and as such should be weighted more heavily if there is a large need throughout the DOE (NRC, 1999d). For example, from Table 2.2, many technology opportunities can be seen for the need to perform the monitoring and characterization of radioactive surfaces and materials during all phases of a D&D project. Therefore, one should expect a large return on investment for science and technology development in the form of reduced characterization and monitoring time.

Step 2. *Prioritize* technology development needs. As part of this step, a comparison of the technologies needed for D&D with the existing available technologies should be made using the criteria from step 1 above. To help accomplish this prioritization uniformly, a set of guidelines should be established and used by all DOE field offices.

Step 3. *Review and evaluate* the technology needs document. As pointed out previously, this is a job for top management that cannot be delegated (NRC, 1998a). The output from this step should be a short list of promising areas for research and technology development.

Defining D&D Task Parameters—An Example for Remote Equipment Needs

To quantitatively define the needs to be addressed in a long-term science program for D&D, a clear definition of the tasks to be performed is essential. Data for this definition can be obtained only by careful analysis of a problem by dedicated, knowledgeable personnel. The existing needs documents fall far short of the numerical clarity that is necessary to plan a two-decade-long research activity. Detailed, quantitative information could be obtained by in-depth analysis of a few of DOE's most significant D&D tasks (i.e., Hanford Building 324, Rocky Flats glove boxes, Oak Ridge Building 3019, and the INEEL Engineering Test Reactor). Data required include a standard project description; a description of distinct physical tasks for that project; planning experience including technology transfer, degree of direct operator involvement, and operator training requirements; deployment issues associated with the proposed technology; and comments on the requirements to drive the needed science. The in-depth analysis should include quantitative information such as the following:

- timeline and frequency of physical tasks;
- duration of each task;
- parameters describing the task (geometry, forces, speeds, accuracy, required dexterity);
- handling requirements (size, shape, weight);
- documentation requirements; and
- radiation levels expected.

It would also be helpful to include data on the D&D task requirements derived from experience with commercial nuclear reactors.

Applying Successful Research Results to DOE's D&D Problem

The OST has recognized an ongoing need to achieve greater integration of basic and applied research programs (DOE, 2000b). A previous study (NRC, 1998a) made a number of recommendations to the D&D focus area (DDFA) aimed at achieving greater deployment of D&D technologies. The Technology Deployment Initiative, Large-Scale Demonstration Projects, and Accelerated Site Technology Deployment Program are examples of efforts to improve the success rate (NRC, 1998a).

Promising EMSP results are unlikely to be moved through the stages of maturity (toward deployment) unless they are sponsored, monitored, and communicated to potential users by the DDFA. Sustained funding at a reasonable level is required to fill the pipeline with science programs and results that are continually beginning as others are completing each year. As noted in the previous section of this report, EMSP funding in the D&D area has been marginal. DOE has identified a significant gap in its environmental quality portfolio relative to research and development (R&D) funding for D&D (DOE, 2000a). Continued funding will be required to generate interest in the scientific community and to instill confidence in the potential users that promising technical areas will be brought to fruition (EMAB, 2001).

In addition to adequate funding and involvement of problem owners in EMSP project selection, a key ingredient in successful application of EMSP results is establishing communication with and involvement of end users at the DOE sites.[5] The recent EMSP workshops (Gilbertson, 2000b) have helped foster such communications, but much more needs to be done. No new technology will be implemented successfully without an end user pulling the rope and actively encouraging and guiding completion of the R&D program. At the time of an EMSP project award, a DDFA representative and a lead end user at one or more sites should be identified. These individuals should receive status reports and have periodic meetings with EMSP-funded scientists and end users at other sites. Funding for travel and labor hours should be provided to the

---

[5]End users are those who will use a given method or technology to accomplish a D&D task. They are usually contractor personnel at DOE sites.

DDFA representative and the end user to enable and encourage these collaborations. Similar incentives have been recommended in a previous study of DOE site cleanup (NRC, 1999f).

A final step in improving application of results is educating the EMSP-funded scientist on the requirements that must be satisfied to implement a new technology. Frequently, the scientist may believe that the project has been completed, when in fact much more is needed to aid the end user in application of the results. For example, to install or apply a new piece of equipment, the end user must have operating procedures, maintenance procedures, and training. The detailed knowledge of the scientist is often needed to prepare these implementation tools. The most effective technology transfers occur when a knowledgeable individual remains involved from the initial research phase through deployment (see Figure 3.1). The EMSP and DDFA should provide funding for continued involvement of the scientists in driving the deployment at operating sites. This would assist the end user during implementation and also give the scientists a greater appreciation of what is required to implement new technology at an operations site. Personnel exchanges between operating sites and research organizations should also be considered.

The committee recommends that OST increase efforts to transition basic research to a deployable product by improving communications and cooperation among the researchers, DOE laboratories, and the contractors performing D&D. Incentives for deployment of newly developed technologies that promise advantages over older technologies should be included in D&D contracts.

Incidental Benefits from the EMSP

In developing the EMSP vision statement requested by Mark Gilberston, the committee found two benefits from EMSP research that are incidental to the EMSP's role as a supplier of research for new EM technologies but are nevertheless important. These include developing future scientists and providing commercially marketable research.

As EM's direct link with the scientific community, the EMSP has an opportunity to enhance its visibility as a developer of new scientists and engineers. In a recommendation on EMSP funding a previous NRC committee stated that "New starts will help establish a cadre of knowledgeable and committed investigators" (NRC, 2000a, p. 127). Similarly the Science Committee of the Environmental Management Advisory Board recommended to EM that the EMSP should build and maintain a cadre of senior scientists and attract younger scientists to work on EM challenges (Berkey, 2000; EMAB, 2001).

Graduate student interest in disciplines needed by the DOE—mathematics, physics, chemical engineering, materials science, mechanical engineering, electrical engineering, and nuclear engineering has decreased over the last ten years (NSF, 1998). Furthermore, there appears to be a lack of student interest in DOE missions and research areas. In many instances undergraduates in the engineering and information technology fields are no longer knowledgeable of DOE laboratories and production facilities due to lack of recent recruitment. The reputation of the laboratories in providing interesting and challenging problems has decreased (DOE, 1999b, 2000c).

In addition to cost savings in DOE's D&D program, technology developed from EMSP research can be expected to provide added value to the national economy. Using information drawn from the Environmental Export Council (EEC), the Environmental Protection Agency has recognized an annual commercial global market of $540 billion for 2000 in environmental technology alone (EPA, 1998). Currently, nearly half of this market is in the United States; in 1996 only 6 percent of the global market was exported to world economies (Mofat and Roa, 1996). The market for environmental technology and science and engineering expertise is broadly acknowledged by the business community and is expected to continue to grow as issues of environmental sustainability and green technology become more engrained in international policy and in public perception (Amato, 2000). Technology developed from fundamental research can also be expected to play a role in developing new products in broader industry segments (e.g., the applications for intelligent remote systems in medicine and the commercial space industry).

References

Amato, I. 2000. Green chemistry proves it pays. Fortune 142(3):270U.

Anderson, G. P., K. D. King, K. L. Gaffney, and L. H. Johnson. 2000. Multi-analyte interrogation using the fiber optic biosensor. Biosensors and Bioelectronics. 14 (10-11):771-77.

Asher, S. A. 2000. Development of Novel, Simple, Multianalyte Sensors for Remote Environmental Analysis. Project Book. EMSP National Workshop 2000. April 24-28. DOE/OSTI-EMSP-1. Washington, DC: U.S. DOE. P. 692.

Banat, I. M., R. S. Makkar, and S. S. Cameotra. 2000. Potential commercial application of microbial surfactants. Applied Microbiological Biotechnology 53:495-508.

Bedick, R. 2000. Overview of the DDFA's International Cooperative Programs. Presentation to the Committee on Long-Term Research Needs for Deactivation and Decommissioning at Department of Energy Sites. Richland, WA. May 24-25, 2000.

Berkey, E. 2000. Committee activities relevant to DOE-EM decontamination and decommissioning. Edgar Berkey, DOE EMSP Advisory Board. Presentation to Academies' Committee on Needs and Opportunities for Deactivation and Decommissioning at Department of Energy Sites. Washington, DC. October 19.

Beuker, H., and F. W. Haesing. 1994. Fiber-optic radiation sensors. Optical Fibre Sensing and Systems in Nuclear Environments. F. Berghmans and M.C. Decreton, eds. Bellingham, WA: The International Society for Optical Engineering 2425:106-14.

Borgermans, P., B. Brichard, F. Berghmans, F. Vos, M. Decréton, K. M. Golant, A. L. Tomashuk, and I. V. Nikolin. 2000. On-line gamma dosimetry with phosphorous and germanium co-doped optical fibres. R. Sharp and P. Dressendorfer, eds. Proceedings of RADECS 1999 Conference. Moscow, Russia. IEEE. 0-7803-5726-4/00. Piscataway, NJ: The Institute of Electrical and Electronics Engineers.

Boyd, G. 2000. Overview of the Environmental Management Clean-up Mission. Presentation to the Committee on Needs and Opportunities for Deactivation and Decommissioning at Department of Energy Sites. Washington, DC. March. 16-17.

Brouns, R. 2000. D&D Research Perspective from the Pacific Northwest National Laboratory. Presentation to the Committee on Needs and Opportunities for Deactivation and Decommissioning at Department of Energy Sites. Richland, WA. May.

Colquhoun, A. P. 1999. Developing, collecting and sharing experience to facilitate planning and implemention of decommissioning programmes. Euradwaste 1999 Conference, Luxembourg. November 15-18. C. Davies, ed. EUR 19173 EN. Luxembourg: Office of Official Publications of the European Communities. P. 129.

Cunningham, A. J. 1998. Introduction to bioanalytical sensors. New York: John Wiley and Sons. P. 417.

Delany, J. M., and S. R. Lundeen. 1990. The LLNL Thermodynamic Database. Technical Report UCRL-21658, Lawrence Livermore National Laboratory. Livermore, California.

DOE (Department of Energy). 1995. Natural and Accelerated Bioremediation Research (NABIR). DOE/ER-0659T. Office of Science. September. Washington, DC: U.S. DOE.

DOE. 1996. Closing the Circle on the Splitting of the Atom, the Environmental Legacy of Nuclear Weapons Production in the United States and What the Department of Energy Is Doing About it. DOE/EM-0266, Office of Environmental Management, January. Washington, DC: U.S. DOE.

DOE. 1997a. Linking Legacies, Connecting the Cold War Nuclear Weapons Production Processes to Their Environmental Consequences. DOE/EM-0319, Office of Environmental Management, January. Washington, DC: U.S. DOE.

DOE. 1998a. Accelerating Cleanup: Paths to Closure. DOE/EM-0362. Washington, DC: U.S. DOE.

DOE. 1998b. Laser Ablation Mass Spectroscopy (LA/MS), Innovative Technology Report, DOE/EM-0369, September. Washington, DC: U.S. DOE.

DOE. 1998c. GammaCam Radiation Imaging System, Innovative Technology Summary Report. OST Reference #1840. February. Washington, DC: U.S. DOE.

DOE. 1998d. Gamma-Ray Imaging System, Innovative Technology Summary Report, DOE/EM-0390. November. Washington, DC: U.S. DOE.

DOE. 1999a. Long Range Alpha Detection for Component Monitoring, Innovative Technology Summary Report. DOE/EM-0497. September. Washington, DC: U.S. DOE.

DOE. 1999b. Maintaining United States Nuclear Weapons Expertise. Chiles Commission report. March. Washington, DC: U.S. DOE.

DOE. 1999c. National Facility Deactivation Initiative. Brochure. Office of Environmental Management. July. Washington, DC: U.S. DOE.

DOE. 2000a. Adequacy Analysis of the Environmental Quality Research and Development Portfolio (Draft), July. Washington, DC: U.S. DOE.

DOE. 2000b. Deactivation and Decommissioning Focus Area Performance Plan. Washington, DC: U.S. DOE.

DOE. 2000c. U.S. Department of Energy Strategic Plan, Strength Through Science Powering the 21st Century (Draft), February 18. Washington, DC: U.S. DOE. P. 71.

DOE. 2000d. Status Report on Paths to Closure. DOE/EM-0526. Washington, DC: U.S. DOE.

DOE. 2000e. Management Plan. Environmental Management, Office of Science and Technology. Research and Development Program. September. Washington, DC: U.S. DOE.

DOE. 2000f. Deactivation and Decommissioning Focus Area Annual Report. FY 2000. DOE/EM-0562. Washington, DC: U.S. DOE.

DOE. 2000g. Environmental Management Science Program. Annual Report. FY 2000. Office of Science and Technology. DOE/EM-0569. Washington, DC: U.S. DOE.

DOE. 2001. Multiyear Program Plan. Fiscal years 2001-2005. Office of Science and Technology Basic and Applied Research. DOE/ID-10840. April.

Dzombak, D. A., and F. M. M. Morel. 1990. Surface Complexation Modeling, Hydrous Ferric Oxide. New York: John Wiley and Sons.

EC (European Commission). 1998. The Commission's Proposal for the 5th Framework Programme (1998-2002). EUR 17651. Luxembourg: Office of Official Publications of the European Communities.

Egge, R. G. 2000. Surveillance/Maintenance and Transition. Presentation to Academies' Committee on Needs and Opportunities for Deactivation and Decommissioning at Department of Energy Sites. Richland, WA: DOE. May 24.

Ehrlich, H. L., and C. L. Brierley, eds. 1990. Microbial Mineral Recovery. New York: McGraw Hill.

EMAB (Environmental Management Advisory Board). 2001. The Role and Status of Basic Science in Accomplishing the Department of Energy's Environmental Management Mission. Environmental Management Advisory Board. Ad Hoc Committee on Science and Innovation. April. Washington, DC: U.S. DOE.

Entine, G., and M. R. Squillante. 1999. Radiation detection. Chapter in Wiley Encyclopedia of Electrical and Electronics Engineering Online. J. Webster, ed. John Wiley & Sons. Available at: <http://www.interscience.wiley.com:83/eeee/eeee_articles_fs.html>.

EPA (Environmental Protection Agency). 1998. EPA Strategy for Promoting U.S. Environmental Exports: A Report to Congress. Washington, DC: U.S. Environmental Protection Agency: Office of International Activities. May. Available at: <http://www.epa.gov/oia/exp420.htm#2>.

Evans, B. D., G. M. Sigel, Jr., J. B. Langworthy, and B. J. Faraday, 1978. The fiber optic dosimeter on the Navigational Technology Satellite 2, IEEE Transactions on Nuclear Science NS-25(6):1619-24.

Forsström, H. 1999. EC R&D: Review of past and on-going research programmes. Euradwaste 1999 Conference, Luxembourg. 15-18. Luxembourg: Office of Official Publications of the European Communities. P. 99.

Fournier, R., P. Gravez, and A. Micaelli. 1998. Computer Aided Teleoperation Recent Trends: The CEA Approaches in Mechanics, Control, and Supervision for Maintenance and Dismantling of Nuclear Plants. IEEE International Conference on Robotics and Automation. Louvain, Belgium. Piscataway, NJ: The Institute of Electrical and Electronics Engineers.

Gilbertson, M. 2000a. Presentation by Mark Gilbertson, Director of DOE Office of Basic and Applied Research. Presentation to Academies' Committee on Needs and Opportunities for Deactivation and Decommissioning at Department of Energy Sites. Washington, DC. March 16-17.

Gilbertson, M. 2000b. Presentation by Mark Gilbertson, Director of DOE Office of Basic and Applied Research. Presentation to Academies' Committee on Needs and Opportunities for Deactivation and Decommissioning at Department of Energy Sites. Washington, DC. October 19-20.

Haley, D. 2000. Deactivation and Decommissioning. Presentation to Academies' Committee on Needs and Opportunities for Deactivation and Decommissioning at Department of Energy Sites. Oak Ridge, TN. June 27.

Hart, P. 2000. Overview of the Deactivation and Decommissioning Focus Area. Presentation to Academies' Committee on Needs and Opportunities for Deactivation and Decommissioning at Department of Energy Sites. Washington, DC. March 17.

Henschel, H., O. Kohn, H. U. Schmidt, E. Bawirzanski, and A. Landers. 1994. Optical fibers for high radiation dose environments. IEEE Transactions on Nuclear Science 41(3):510-16.

Hochel, R. C. 2000. New Measurement Techniques and Results for Tritium in Concrete. Presentation at 220th American Chemical Society National Meeting in Washington, DC. August 20-24.

Holmes, M. H. 2000. Fort St. Vrain Reactor Decommissioning. Presentation by M. H. Holmes, Utility Engineering Company, to Academies' Committee on Needs and Opportunities for Deactivation and Decommissioning at Department of Energy Sites. Westminster, CO. August 24.

Holtz, J. H., and S. A. Asher. 1997. Polymerized colloidal crystalline hydrogel films as intelligent chemical sensing materials. Nature 389:829-32.

Huston, A. L., B. L. Justus, and T. L. Johnson. 1996. Fiber-optic-coupled, laser heated thermoluminescence dosimeter for remote radiation sensing. Applied Physics Letters 68:3377.

IAEA (International Atomic Energy Agency). 1999. State of the Art Technology for Decontamination and Dismantling of Nuclear Facilities. Technical Reports Series No. 395. Vienna: IAEA.

ISTC (International Science and Technology Center). 2001. Available at: <http://www.istc.ru>.

Jager, E. W. H., O. Inganäs, and I. Lundström. 2000. Microbots for Micrometer-Size Objects in Aqueous Media: Potential Tools for Single-Cell Manipulation. Science. 228:2335. June 30.

Janata, J. 1989. Principles of Chemical Sensors. Modern Analytical Chemistry. New York: Plenum.

Joannopoulos, J. D., R. D. Meade, and J. N. Winn. 1995. Photonic Crystals: Molding the Flow of Light. Princeton, NJ: Princeton University Press.

Justus, B. L., K. J. Pawlovich, C. D. Merritt and A. L. Huston. 1999a. Optically and Thermally Stimulated Luminescence Characteristics of Cu^{1+}-Doped Fused Quartz. Radiation Protection Dosimetry. 81(1):5-10.

Justus, B. L., C. D. Merritt, K. J. Pawlovich, A. L. Huston and S. Rychnovsky. 1999b. Optically stimulated luminescence dosimetry using doped fused quartz. Radiation Protection Dosimetry. 84(1-4):189-92.

Kress-Rogers, E., ed. 1997. Handbook of Biosensors and Electronic Noses: Medicine, Food and the Environment, p. 695. Boca Raton, Fla.: CRC Press.

Lee, C. W., J. Min, S. H. Moon, R. A. LaRossa, and M. B. Gu. 2000. Detection of radiation effects using recombinant bioluminescent Escherichia coli strains. Radiation Environment. Biophysics 39:41-45.

Lieberman, R. A., ed. 1996. Chemical, biochemical and environmental fiber sensors VIII. Proceedings of SPIE. 2836. Bellingham, WA: Society of Photo-optical Instrumentation Engineers.

Lloyd, J. E., J. S. Beis, D. D. Pai, and D. G. Lowe. 1999. Programming Contact Tasks Using a Reality-based Environment Integrated with Vision. IEEE Transactions on Robotics and Automation. 15(3):423-34.

MARSSIM. 2000. Multi-agency radiation survey and site investigation manual (MARSSIM), DOE/EH-0624 Rev. 1, NUREG-1575 Rev. 1, EPA 402-R-97-016 Rev. 1, August. Washington, DC: Government Printing Office.

MacFarlan, G. 2000. Hanford 200-Area Canyon Disposition Initiative. Presentation to Academies' Committee on Needs and Opportunities for Deactivation and Decommissioning at Department of Energy Sites. Richland, WA. May 24.

MacLachlan, A. 2000. Radiation experts don't agree on how to release materials. Nucleonics Week 41:13.

McDaniel, L. 2000. Spent Nuclear Fuels Facility Deactivation. Presentation to Academies' Committee on Needs and Opportunities for Deactivation and Decommissioning at Department of Energy Sites. Richland, WA. May 24.

Mecklenburg, M., B. Danielsson, H. Boije, I. Surugui, and B. Rees. 2000. DNA-Based Biosenors: A Tool for Environmental Analysis. In: A. Mulchandani and O. A. Sadik, eds. Chemical and Biological Sensors for Environmental Monitoring. Washington, DC: American Chemical Society. P. 310.

Menon, S. 2000. International D&D Research Perspective. Presentation to the Committee on Long-Term Research Needs for Deactivation and Decommissioning at Department of Energy Sites. Oak Ridge, TN. June 26-27, 2000.

Meservey, R. 2000. Research Needs for Future Decommissioning Activities at the Idaho National Engineering and Environmental Laboratory. Presentation to Academies' Committee on Needs and Opportunities for Deactivation and Decommissioning at Department of Energy Sites. Richland, WA. May.

Miller, D. J., and R. C. Lennox. 1991. An Object-Oriented Environment for Robot System Architectures. IEEE Control Systems Magazine. Pp. 14-23. February.

Miller, R. W., A. Huston, B. Justus, H. Ning, S. Worely, T. Bevels, R. Altemus, and C. N. Coleman. 2000. Dosimetry characteristics of an optically stimulated luminescent fiberoptic dosimetry system. Paper presented at the 2000 World Congress on Medical Physics and Biomedical Engineering, Chicago, IL. July 23-28.

Mofat, S., and R. M. Roa. 1996. Asia stink. Fortune. December 9. Available at: <http://www.fortune.com/fortune/1996/961209/big.html>.

Moulin, A. M., S. J. O'Shea, and M. E. Welland. 2000. Microcantilever-based biosensors. Ultramicroscopy. 82:23-31.

NATO (North Atlantic Treaty Organization). 2000a. The NATO Science Programme. Available at: <http://www.nato.int/science/index.html>.

NATO. 2000b. The challenge of modern society. Available at: <http://www.nato.int/ccms/index.html>.

NCES. 1995. National Center for Education Statistics. Earned Degrees and Completion Surveys. Washington, DC: NCES.

NEA (Nuclear Energy Agency). 1996. The First Ten Years 1985-95: The NEA Co-operative Programme on Decommissioning. Paris: OECD/NEA.

NEA. 1998. Mission Statement, Strategic Options, Values: Co-operative Programme for the exchange of scientific and technical information concerning nuclear installation decommissioning projects, NEA/CPD/DOC(98)4, August.

NRC (National Research Council). 1996. Affordable Cleanup? Opportunities for Cost Reduction in the Decontamination and Decommissioning of the Nation's Uranium Enrichment Facilities. Washington, DC: National Academy Press.

NRC. 1997. Building an Effective Environmental Management Science Program: Final Assessment. Washington, DC: National Academy Press.

NRC. 1998a. A Review of Decontamination and Decommissioning Technology Development Programs at the DOE. Washington, DC: National Academy Press.

NRC. 1998b. Issues in Potable Reuse: The Viability of Augmenting Drinking Water Supplies with Reclaimed Water. Washington, DC: National Academy Press.

NRC. 1999a. An End State Methodology for Identifying Technology Needs for Environmental Management, with an Example from the Hanford Site Tanks. Washington, DC: National Academy Press.

NRC. 1999b. The State of Development of Waste Forms for Mixed Wastes. Washington, DC: National Academy Press.

NRC. 1999c. Technologies for Environmental Management. Washington, DC: National Academy Press.

NRC. 1999d. Improving Project Management in the Department of Energy. Washington, DC: National Academy Press.

NRC. 1999e. Decision Making in the U.S. Department of Energy's Environmental Management Office of Science and Technology. Washington, DC: National Academy Press.

NRC. 1999f. Groundwater and Soil Cleanup: Improving the Management of Persistent Contaminants. Washington, DC: National Academy Press.

NRC. 2000a. Research Needs in Subsurface Science. Washington, DC: National Academy Press.

NRC. 2000b. Long-Term Institutional Management of U.S. Department of Energy Legacy Waste Sites. Washington, DC: National Academy Press.

NRC. 2001a. Research Needs for Managing High-Level Waste in the Department of Energy. Washington, DC: National Academy Press.

NRC. 2001b. A Strategic Vision for Department of Energy Environmental Quality Research and Development. Washington, DC: National Academy Press.

NSF (National Science Foundation). 1998. Science Resources Studies Division, Graduate Students and Postdoctorates in Science and Engineering, Science and Engineering Indicators 1998. Available at: <http://www.nsf.gov>.

NSF. 1999. Nanotechnology—Shaping the World Atom by Atom (NSTC report). I. Amato. 1999. Available at: <http://itri.loyola.edu/nano/IWGN.Public.Brochure/>.

NSTC. 1999. Nanotechnology Research Directions: IWGN Workshop Report. National Science and Technology Council. Interagency Working Group on Nanoscience, Engineering and Technology, Washington, DC, September. Available at: <http://itri.loyola.edu/nano/IWGN.Research.Directions/>.

Nuclear Waste News. 2000. Metals recycling: DOE releases new directives on recycling surplus metal. 20 (41): 403-04. October 12.

Parker, F. L., K. W. Ayers, et al. 1998. Reuse of Concrete from Contaminated Structures. Proceedings of Spectrum '98. Denver, Colo. September 13-18. La Grange Park, IL: American Nuclear Society. 1:71-74.

Rogers, R. D., M. A. Hamilton, L. O. Nelson, J. Benson, and M. Green. 1997. Evaluation of Microbially Influenced Degradation as a Method for the Decontamination of Radioactively Contaminated Concrete. Materials Research Society Conference Proceedings. 0465:0317.

Sandia. 1998. Robotics and Intelligent Machines in the U.S. Department of Energy. A Critical Technology Roadmap. October. Sandia National Laboratory Report SAND98-2401/2.

Simpson, M. L., G. S. Sayler, B. M. Applegate, S. Ripp, D. E. Nivens, M. J. Paulus, and G. E. Jellison Jr. 1998. Bioluminescent-bioreporter integrated circuits form novel whole-cell biosensors. Trends in Biotechnology 16:332-38.

Squillante, M. R., and K. Shah. 1995. Other Materials: Status and Prospects, in T. E. Schlessinger and J. B. James (eds.), Semiconductors for Room Temperature Detector Applications. Semiconductors and Semimetals: Nonlinear Optics in Semiconductors, vol. 43. San Diego: Academic Press.

Stewart, D. B., R. A. Volpe, and P. K. Khosla. 1992. Integration of Real-Time Software Modules for Reconfigurable Sensor-Based Control Systems. Proceedings of IEEE/RSJ International Conference on Intelligent Robots and Systems. Piscataway, NJ: The Institute of Electrical and Electronics Engineers. Pp. 325-32.

Sturzenbecker, M. C. 1991. Building an Object-Oriented Environment for Distributed Manufacturing Software. Proceedings of IEEE International Conference on Robotics and Automation. Piscataway, NJ: The Institute of Electrical and Electronics Engineers. Pp. 1972-8.

Sullivan, E. R. 1998. Molecular genetics of biosurfactant production. Current Opinion in Biotechnology 62:1-34.

Thundat, T. G., K. B. Jacobsen, M. J. Doktycz, S. J. Kennel, and R. J. Wamack. 2000. Micromechanical antibody sensor. Patent Cooperation Treaty, International Patent Application (Patent: CODEN:PIXXD2 WO0058729).

Tomashuk, A. L., K. M. Golant, E. M. Dianov, I. V. Nikolin, and R. R. Khrapko. 1999. Principle of operation of fiber optic dosimeter. Electronics Letters 35(2):170-71.

Torma, A. E. 1988. Leaching of metals. In: H. J. Rehm and G. Reed, eds. Biotechnology. Weinheim, Germany: VCH Verlagsgesellschaft. Pp. 367-99.

Updike, S. J., and G. P. Hicks. 1967. The enzyme electrode. Nature 214:986.

van der Lee, J., E. Ledoux, G. de Marsily, A. Vinsot, van de Weerd, A. Leijnse, B. Harmand, E. Rodier, M. Sardin, J. Dodds, and A. Hérnandez Benitez. 1997. Development of a Model for Radionuclide Transport by Colloids in the Geosphere. Nuclear Science and Technology. EUR 17480 EN. Brussels, Belgium: European Commission.

Van Hecke, G. R., and K. K. Karukstis. 1998. A Guide to Lasers in Chemistry. Boston, MA: Jones and Bartlett Publishers.

Webster, J. G., ed. 1999. The Measurement, Instrumentation and Sensors Handbook. Boca Raton, FL: CRC Press.

Young, L. L., and C. E. Cerniglia, eds. 1995. Microbial Transformation and Degradation of Toxic Organic Chemicals. New York: John Wiley and Sons. P. 654.

Appendixes

A
Presentations to the Committee

Washington, D.C., March 16-17, 2000

Overview of the Environmental Management Cleanup Mission, Gerald Boyd, DOE Office of Environmental Management (EM)
Description of the EM Science Program (EMSP), Roland Hirsch, DOE Office of Energy Research (ER), and Mark Gilbertson, DOE-EM
Statement of Task and Needs of the Sponsor, Mark Gilbertson, DOE-EM
Overview of the Deactivation and Decommissioning Focus Area (DDFA), Paul Hart, National Energy Technology Laboratory (NETL)
Site Needs and Research Opportunities, Paul Hart, NETL
Program Coordination, Mary McCune, DOE-EM

Richland, Washington, May 24-25, 2000

Presentations during the Hanford Site tour:
General Site Information, Kim Koegler, Bechtel Hanford
Site Decommissioning Projects, Mark Morton, Bechtel Hanford
K-Basins, Larry McDaniel, Bechtel Fluor Hanford
B-Reactor, B Reactor Museum Association
Surveillance/Maintenance and Transition Projects, Bob Egge, Bechtel Hanford
Canyon Disposition Initiative and 221-U, Gary MacFarlan, Bechtel Hanford
Plutonium Finishing Plant, Mark Gibson, Bechtel Fluor Hanford
233-S Plutonium Concentration Facility Decommissioning Project, Allan Chaloupka, Bechtel Hanford
324 Hot Cells, Tim Erickson, Bechtel Fluor Hanford

Presentations at the hotel meeting room:

Future Deactivation and Decommissioning (D&D) Challenges at Hanford, John Sands, DOE-Richland Operations Office

Brownfield Plans for the 300 Area, Jay Augustenborg, DOE-Richland Operations Office

D&D Programs and Perspectives at the Pacific Northwest National Laboratory (PNNL), Richard Brouns, PNNL

Future D&D Challenges at the Idaho National Engineering and Environmental Laboratory (INEEL), Dick Meservy, INEEL

Future D&D Challenges at the Savannah River Site, Ed Stevens, Savannah River Technology Center

Perspectives and Concerns of the Hanford Advisory Board (HAB), Harold Heacock, HAB

D&D Summary from the National EMSP Workshop, Bob Bedick, NETL

Oak Ridge, Tennessee, June 26-27, 2000

Presentations during the Oak Ridge Site tour:

East Tennessee Technology Park and K-33 Building, Jack Howard, DOE-Oak Ridge Operations Office; Jim McAnally, Danny Nichols, and Ricky Miles, BNFL, Inc., Oak Ridge

Oak Ridge National Laboratory (ORNL) Research Reactor, Jim Blair, Bechtel-Jacobs

Y-12 Isotope Separations, Don Sparkman, Bechtel-Jacobs

Presentations at hotel meeting room:

ORNL D&D Program, Jim Blair, Bechtel-Jacobs

East Tennessee Technology Park D&D Overview, Gary Person, Bechtel-Jacobs

Oak Ridge Operations Site Technology Coordinating Group, Paula Kirk, Bechtel-Jacobs

Views from the DDFA Lead Laboratory, Dennis Haley, ORNL

Robotics and Remote Systems, Dennis Haley, ORNL

Views of the Oak Ridge Citizens Advisory Board, William Pardue, Oak Ridge Site Specific Advisory Board

International D&D Research Perspective, Shankar Menon, Nyköping, Sweden

Westminster, Colorado, August 23-24, 2000

Presentations at the Rocky Flats visitors center:
Rocky Flats Closure Overview, Tom Lukow, DOE-Rocky Flats Field Office
Rocky Flats Closure Baseline, Nancy Tuor, Kaiser-Hill
Decommissioning and Demolition Technology Deployment, Jeff Stevens, Kaiser-Hill
Comments by Rocky Flats Citizens Advisory Board, Gerald DePoorter, Rocky Flats Citizens Advisory Board

Presentation during Rocky Flats Environmental Technology Site tour:
Former Plutonium Recovery Building 771, Joe Springer, DOE-Rocky Flats Field Office

Presentation at the hotel meeting room:
Fort St. Vrain Reactor Decommissioning, Mike Holmes, Utility Engineering

Washington, D.C., October 19, 2000

U.S. Nuclear Regulatory Commission (USNRC) Programs Relevant to D&D, Bill Ott, USNRC
Overview of DOE Environmental Management Advisory Board (EMAB) Functions Relevant to EMSP D&D, Ed Berkey, EMAB
DOE Office of Science Low Dose Program, David Thomassen DOE-SC
Comments from the Sponsor, Mark Gilbertson, DOE-EM

B

Biographical Sketches of Committee Members

PHILIP R. CLARK (NAE) (Chair) is a 45-year veteran in the field of naval reactors and commercial nuclear power. From 1950 to his retirement in December 1995, he was with GPU Nuclear Corporation of which he was President and CEO from 1983-95. There he was responsible for cleanup of the damaged TMI-2 plant, restarting TMI-1, operating Oyster Creek, and decommissioning Saxton. As a manager he has been responsible for research, development, design, construction, operation, maintenance, training, quality assurance, budget, and decommissioning cleanup. Mr. Clark is a member of the National Academy of Engineering and a fellow of the American Nuclear Society. During his government service, he was associate director of reactors in the Naval Reactor Division of the U.S. Department of Energy and chief of the Reactor Engineering Division, Nuclear Power Directorate, Naval Sea Systems Command. While in these positions he directed a major element of the U.S. Naval Nuclear Propulsion Program and received the Navy Distinguished Civilian Service Award and the U.S. Energy Research and Development Administration Special Achievement Award. Mr. Clark earned a B.S. degree in civil engineering in 1951 from Polytechnic Institute of Brooklyn, where he has also done graduate study. Mr. Clark has served on several NRC committees, including the Committee on Improving Project Management, the Committee on Decontamination and Decommissioning of Uranium Enrichment Facilities, and the NAE Membership Task Group.

ANTHONY CAMPILLO is the head of the Optical Physics Branch at the Naval Research Laboratory, where he directs research in the areas of laser development, optical diagnostic and sensor development, photonics, and nanotechnology. He is an active researcher who currently publishes in the areas of laser chemical analysis, photonic bandgap structures, and microcavity effects. He has also done research in laser shock generation, quantum dot optics, nanofabrication, optical limiters for eye and sensor protection, and bio-aerosol detection and characterization. Dr. Campillo has authored over 100 refereed journal articles and 1 book, presented 150 talks at professional society meetings and has 8 patents issued or pending. He has been a member of the technical staffs of General Telephone & Electronics Laboratory, Los Alamos Scientific Laboratory and Brookhaven National Laboratory. Dr. Campillo is a fellow of the Optical Society of America and the American Physical Society and is active in the Laser Electro-optics Society of the IEEE and the American Association of Aerosol Research. He is the associate editor of the journal *Optics Letters*. Dr. Campillo's formal training is in electrical engineering with a B.S. degree from New Jersey Institute of Technology, an M.S. degree from Princeton University, and a Ph.D. degree from Cornell University.

FRANK CRIMI is a retired vice president of Lockheed Martin Advanced Environmental Systems Company. He joined Lockheed after completing 34 years with the General Electric Company. Mr. Crimi has over 43 years of experience in the design, operations, and maintenance of nuclear power plants with emphasis on the decontamination and decommissioning of nuclear facilities. His experience includes management of large, complex programs in the nuclear industry and construction, operation, and maintenance of naval nuclear power plants. He managed General Electric's Decommissioning Services and was the General Electric program manager for the decontamination and decommissioning of the Shippingport Atomic Power Station. Mr. Crimi has served as a member of several National Research Council committees reviewing the cost to decommission DOE's Uranium Enrichment Facilities, DOE's decontamination and decommissioning technology development program, DOE's peer review process, and DOE's policies and practices to design, procure, and manage major construction projects. Currently he is serving as a member of the committee reviewing and evaluating the Army's Chemical Stockpile Disposal Program. Mr. Crimi holds a B.S. degree in mechanical engineering from Ohio University and has done graduate work in mechanical engineering at Union College, Schenectady, New York. He is a member of the American Society of Mechanical Engineers and American Nuclear Society and has written and presented several papers on the decommissioning of nuclear facilities.

KEN CZERWINSKI is an assistant professor in the Department of Nuclear Engineering at the Massachusetts Institute of Technology and a consultant for the Idaho National Engineering and Environmental Laboratory. His expertise is in actinide thermodynamics, environmental chemistry of the actinide elements, development of actinide-transactinide separations, and synthesis of inorganic complexation agents. Dr. Czerwinski has been an associate research scientist for the Institut für Radiochemie Technische Universität München and a postdoctoral fellow of the Glenn T. Seaborg Institute for Transactinium Science at the Lawrence Livermore National Laboratory. He has been accorded the Department of Energy's Defense Programs Early Career Scientist and Engineer Award and the Carl Richard Soderberg Professorship in Power Engineering. He is currently chair of the Norman C. Rasmussen Career Development and a consultant for the INEEL's migration of plutonium from transuranic waste areas project. Dr. Czerwinski earned his B.A. degrees in chemistry and the Russian language from Knox College in Galesburg, Illinois, and his Ph.D. in nuclear chemistry from the University of California, Berkeley.

RACHEL DETWILER is principal engineer of materials research and consulting for Construction Technology Laboratories, Inc., in Skokie, Illinois. Her area of expertise is in the durability of concrete and cement-based materials used for storage of radioactive wastes. Dr. Detwiler has been involved in designing and testing a grout for the stabilization of radioactive and hazardous waste in underground storage tanks at the Savannah River Site and has participated in a range of tests on the durability of concrete. She has served as an assistant professor at the University of Toronto; postdoctoral research fellow at Norges Tekniske Hogskole, Trondheim, Norway; and design and materials engineer with ABAM Engineers, Inc. She is a member of the American Society for Testing and Materials and the American Concrete Institute where she has served as chair of Committee 227 on Radioactive and Hazardous Waste Management and as a member of Committee 234 on Silica Fume in Concrete. She has received a Norges Teknisk-Naturvitenskapelige Forskningrad Fellowship and the Carlson-Polivka Fellowship and has written over 30 technical papers related to concrete microscopy, durability, and testing. Dr. Detwiler earned a B.S. degree in civil engineering, an M.S. degree in structural engineering, and her Ph.D. in civil engineering materials, all from the University of California, Berkeley.

HARRY HARMON is a senior program manager for Pacific Northwest National Laboratory. His expertise is in waste management, nuclear processing, separations chemistry and engineering, and implementing environmental programs and developing technology in these and related areas. His experience includes managing NUKEM's waste management and environmental remediation activities; leading tank waste programs as the vice-president of tank waste programs at M4 Environmental Management, Inc.; serving as technical director for all high-level waste technology development at the Westinghouse Savannah River Company; serving as Vice President of Tank Waste Remediation System Division of Westinghouse Hanford Company—the overall system required to safely manage the waste tanks and process the waste for disposal; and managing process and equipment research and development at the Savannah River Laboratory. He is a member of the American Chemical Society and Sigma Xi. He has also written or co-written a collection of articles and publications. Dr. Harmon earned a B.S. degree in chemistry from Carson-Newman College, Jefferson City, Tennessee and a Ph.D. in inorganic and nuclear chemistry from the University of Tennessee, Knoxville.

VINCENT MASSAUT is a project manager at the National Research Center (SCK-CEN) in Mol, Belgium, and a lecturer at the University of Liège. His expertise is in decommissioning nuclear installations. Since 1989 he has been involved in research for the pilot BR-3 reactor decommissioning project, first as an assistant to the project leader, then as project manager. He developed the first decommissioning plan for a pressurized water reactor in Belgium. He is a member of European Commission's (EC's) group of experts on D&D and waste management issues for the EC's Joint Research Centres, a member of the EC dismantling working group, and a member of the cooperative agreement on decommissioning of the OECD/NEA. Previously, Mr. Massaut was coordinator of decommissioning projects for the third and fourth 5-year programs of the EC and a member of the Decommissioning Steering Group aimed at helping the EC in defining its future policy related to R&D in decommissioning. He was selected as an IAEA expert to advise on dismantling problems, decontamination, and site restoration in Lithuania and Poland. He earned a M.S. degree in engineering with a specialization in nuclear science from the University of Liège.

ALAN PENSE (NAE) is professor emeritus at the ATLSS Engineering Research Center of Lehigh University in Pennsylvania. As a specialist in physical and mechanical metallurgy he has taught undergraduate and graduate courses in general metallurgy, heat treatment, fracture, failure analysis, and failure prevention. Dr. Pense has conducted sponsored research on the properties and welding of materials for the American Iron and Steel Institute, the National Science Foundation, the Welding Research Council, the National Cooperative Highway Research Board, the U.S. Department of Transportation, and the Pennsylvania Department of Transportation. His expertise focuses on the practical application of steel metallurgy and welding technology to large structures and structural systems, including pressure vessels, bridges, buildings, nuclear components, and ships. This research has resulted in a volume of published work, and three of his papers have won awards from the American Welding Society. Dr. Pense consults for private industry and government agencies, including the Nuclear Regulatory Commission, major steel companies, state departments of transportation, and design and construction firms. He has received the Robinson Award from Lehigh University, the Student Council Teaching Award, the Stabler Teaching Award, the American Society for Engineering Education Western Electric Teaching Award, and the William Hobart Medal and has been elected to the National Academy of Engineering. He has authored or co-authored some 100 articles and book chapters. Dr. Pense earned his B.S. degree in metallurgical engineering from Cornell University and his M.S. and Ph.D. degrees from Lehigh University.

GARY S. SAYLER is a professor of microbiology and ecology and evolutionary biology at the University of Tennessee. He received his Ph.D. in bacteriology and biochemistry from the University of Idaho, where he conducted research on heterotrophic turnover of organic matter in freshwater environments. This was followed by postdoctoral training in marine microbiology and biodegradation at the University of Maryland. He is the founding director of the Center for Environmental Biotechnology and director of the State Center of Excellence, the Waste Management Research and Education Institute. He has edited five books and contributed to 215 publications in broad areas of environmental microbiology, biodegradation, and biotechnology and holds patents on environmental gene probing, genetic engineering for bioremediation, and bioelectronic sensor technology. His work has included molecular and environmental aspects of PCB, PAH, BTEX, and TCE biodegradation. He has served on numerous panels and chaired advisory review committees of ORNL, NIH, DOE, EPA, and others. During his career, he has received the American Society for Microbiology Procter and Gamble Award for Environmental Microbiology, the Distinguished Alumni Award of the University of Idaho, and the Dow Chemical Foundation SPERE Award. He has been elected to the American Academy of Microbiology. He has served in an editorial capacity for six journals and is currently an editor for *Environmental Science and Technology*. Professional memberships include American Association for the Advancement of Science, American Chemical Society, and a few others. Dr. Sayler is a member of the Water Environment Research Foundation, Research Council.

DELBERT TESAR holds the Carol Cockrell Curran Chair in Engineering and is professor of mechanical engineering at the University of Texas at Austin and is director of the Robotics Research Group there. His expertise is in robotics, intelligent mechanical systems, and the interactive design of such systems. He has pursued research in the machine systems field for 40 years and has graduated 45 Ph.D.'s and 125 M.Sc.s with theses. Recent work has concentrated on a full architecture for intelligent actuators, which as standard building blocks can be assembled on demand into an open architecture system (from 10 to 40 degrees of freedom), all operated by one universal software system. He has been a member of the Air Force Review Committee for the MANTECH Program, the Air Force Science Advisory Board, three national review panels on robotics for National Institute of Science and Technology and NASA, and served on a standing review committee of the National Research Council on the space station. Dr. Tesar earned both his B.S. degree in mechanical engineering and his M.S. degree in engineering mechanics from the University of Nebraska and his Ph.D. in mechanical engineering from the Georgia Institute of Technology.

C
Interim Report

THE NATIONAL ACADEMIES
Advisers to the Nation on Science, Engineering, and Medicine

National Academy of Sciences
National Academy of Engineering
Institute of Medicine
National Research Council

Board on Radioactive Waste Management
National Research Council

December 5, 2000

Dr. Carolyn Huntoon
Assistant Secretary for Environmental Management
U.S. Department of Energy
Washington, DC 20585

Dear Dr. Huntoon:

    At the request of the Department of Energy's (DOE's) Office of Environmental Management, the National Research Council (NRC) empaneled a committee[1] to assist the Department in developing a long-range science plan for deactivation and decommissioning (D&D) research sponsored by the Environmental Management Science Program (EMSP).[2] The EMSP requested that the committee write an interim report primarily to provide advice on the technical content of a fiscal year 2001 (FY 01) EMSP call for research proposals[3] and to give other advice the committee believes would be helpful. Accordingly, the committee directed most of its efforts toward the second item of its task statement: "recommend areas of research where the EM Science Program can make significant contributions to solving [D&D] problems and adding to scientific knowledge generally." Some general advice, which touches on other portions of the task statement and which will be developed further in the final report, is given in the last section of this interim report. This letter provides the requested interim report, which reflects a consensus of the committee and has been reviewed in accordance with the procedures of the NRC.[4]

    The information used to develop this interim report was obtained from several sources.[5] The committee reviewed previous NRC reports relevant to this study (NRC, 1996, 1997, 1998, 1999b). The committee also held four information-gathering meetings to familiarize itself with D&D challenges throughout the DOE complex.[6] The first meeting, which was held on March 16-17, 2000 in Washington, D.C., provided the committee with an overview of the Office of Environmental Management's (EM's) plans for site cleanup, the EMSP, and the activities of the

---

[1] Committee on Long-Term Research Needs for Deactivation and Decommissioning at Department of Energy Sites. The roster for this committee is given in Attachment A.
[2] The committee's statement of task is given in Attachment B.
[3] DOE intends to publish the proposal call in the Federal Register in December 2000 for funding in FY 01.
[4] The list of reviewers is given in Attachment C.
[5] The reference list is in Attachment D.
[6] DOE has defined its D&D task to include buildings and equipment therein: production and research reactors; facilities for chemical processing, uranium and plutonium processing, and tritium extraction; and gaseous diffusion plants.

Dr. Carolyn Huntoon
December 5, 2000
Page 2

D&D Focus Area (DDFA) within EM's Office of Science and Technology (OST). Three subsequent meetings, which included tours of facilities that are illustrative of major D&D challenges, were arranged as follows:

- Hanford, May 24-25, 2000, focused on production reactor, separations, and fuel fabrication facilities;
- Oak Ridge, June 26-27, 2000, focused on research reactor, gaseous diffusion, and laboratory facilities; and
- Rocky Flats, August 23-24, 2000, focused on plutonium handling facilities and lessons learned.

The committee has largely completed its fact finding activities and has had significant discussion of findings and recommendations. While further discussions will refine and expand on this interim report, the committee offers the following comments that it believes will be helpful in preparing the FY 01 solicitation and in developing plans for a long-term program.

The committee finds that there are strong safety and economic incentives for innovative D&D technologies that may be achieved through scientific research.

The safety incentive is immediate for workers conducting D&D operations, and it will grow as DOE takes on the more challenging D&D tasks. These workers deal with special hazards that are different from those in other parts of DOE's Paths to Closure program (USDOE, 1998a), including the following:

- working in confined spaces in areas of high radioactivity,
- disassembling and removing massive steel and concrete structures,
- direct, hands-on manual labor with powerful saws, torches, and lifting devices, and
- incomplete knowledge of the highly complex systems they are dismantling.[7]

DOE expects to spend some $30 billion for D&D of weapons complex facilities after 2006, compared to about $4 billion until then (Hart, 2000). This is because the biggest D&D challenges, for example at the Savannah River and Hanford sites, will be undertaken after 2006. The DDFA believes that about half of the $30 billion can be saved through use of innovative technologies that it expects could be developed by that time (Hart, 2000).

RECOMMENDATIONS FOR THE EMSP FY 01 SOLICITATION

To meet EMSP's needs for its upcoming solicitation, the committee has identified, preliminarily, three areas where it feels present technology is inadequate and where it believes EMSP-funded research could make significant contributions. These areas are characterization, decontamination, and remote systems. Within these three areas the committee has five specific recommendations that EMSP may wish to consider in preparing its forthcoming solicitation. Two recommendations deal with characterization, two deal with decontamination, and one deals with the crosscutting area of remote systems (including robotics).

---

[7] For example, it is not uncommon for workers to encounter toxic or radioactive materials trapped in unexpected places in pipes or ductwork.

Dr. Carolyn Huntoon
December 5, 2000
Page 3

Characterization

Characterization of contaminated materials is critical at several stages of D&D. Initially, the nature and extent of contamination with both radionuclides and toxic materials must be accurately assessed to ensure adequate protection of workers and the environment, as well as to allow the selection of appropriate methods of decontamination. During decontamination and/or demolition of contaminated equipment and structures, there must be some means of monitoring progress and potential contaminant releases. Finally, after decontamination, the nature and extent of residual contamination must be assessed to determine the final classification and disposal of the item in question.

(1) The committee recommends basic research toward identification and development of means, preferably real-time,[8] minimally invasive, and field usable, to locate and quantify difficult to measure contaminants significant to D&D. These means should be applicable to the major materials and configurations of interest, such as concrete, stainless steel, and packaged wastes. The solicitation should identify the contaminants of interest, including tritium, technetium-99, plutonium-239 and other actinides, beryllium, mercury, asbestos, and polychlorinated biphenyls (PCBs). The need for such characterization was recognized in several presentations to the committee during its site visits, but there are no currently funded EMSP D&D projects in this area.

Rationale: The varied nature of D&D facilities has led to a wide range of contaminant types and site-specific characterization challenges, each generally requiring a detector (Janata, 1989; Webster, 1999;) tailored specifically to the contaminant being measured and its matrix. Some 2,700 buildings, constructed mostly of concrete and containing 180,000 metric tons of metals, are currently within EM's D&D task (Hart, 2000). In its fact finding the committee identified four broad areas where research can advance the state of art: (1) methods to assess the distribution of contaminants within concrete; (2) sensors to measure contaminants on the surface and within micro-cracks of metals; (3) remote sensing of contaminants; and (4) biosensors (see next recommendation).

The development of minimally- and non-invasive real-time *in situ* sensing technologies to characterize the concentration of contaminants, as a function of depth within concrete, would eliminate difficulties associated with core sample collection and subsequent analysis. Minimally invasive schemes like laser ablation mass spectroscopy (Van Hecke and Karukstis, 1998) or non-intrusive techniques like neutron activation and x-ray analysis appear to be attractive candidates for further research.

More sensitive detectors,[9] for example for alpha particles (USDOE, 1999), as well as simple-to-use techniques, such as chemical indicators (Holtz and Asher, 1997), are needed to quickly certify levels of nuclides, hazardous metals, and other toxic substances on structural surfaces and equipment. This will help ensure safety in the workplace and reduce costs—for example by allowing non-hazardous waste to be disposed in landfills. Analysis of residual low-energy beta emitters like tritium and Tc-99 is particularly challenging when these isotopes are

---

[8] Real-time characterization would provide information to workers as they performed a task. For example, as they decontaminate a surface or sort waste.

[9] See, for example, Chapter IX, "Radiation Measurement" of Webster, 1999.

Dr. Carolyn Huntoon
December 5, 2000
Page 4

inside equipment or mixed in heterogeneous waste matrices, because the beta articles cannot penetrate through most materials.

Remote sensing systems can provide both economic and safety benefits by distancing the worker from hazardous work areas. Remote mapping of activity levels using gamma cameras (USDOE, 1998b) is now being used to great advantage in D&D operations. Smaller, higher sensitivity and resolution versions of these instruments would be desirable and may be achievable through further research on detector materials and geometries. Fiber-optic sensing for remote detection of some chemical species is feasible (Lieberman, 1996). Further research could lead to its use in sensing chemical contaminants relevant to D&D. Fiber-optic radiation sensors are a more recent development (Henschel et al., 1992; Huston et al., 1996; Borgermans et al., 2000;), and opportunities exist for both improved performance and novel features such as optical interrogation (Miller et al., 2000).

(2) The committee recommends basic research that could lead to development of biotechnological sensors to detect contaminants of interest. Such research may provide a completely new way to meet the needs for characterization of contaminated materials, which were described in the earlier parts of this section. We believe specific attention to this research opportunity is warranted because the field of biotechnology is rapidly expanding, its potential was not recognized in site presentations heard by the committee, and no biosensor project was funded in the EMSP's previous D&D research solicitation. The contaminants of interest and the materials and configurations in which they must be detected, noted in (1), should be specified in the EMSP's request for proposals.

Rationale: Biosensor technology is a relatively young field originating in 1967 (Updike and Hicks, 1967). However, there has been tremendous growth in development and commercialization of a broad range of biosensor devices and applications (Kress-Rogers, 1997). Modern devices can range from fiber-optic (Anderson et al., 2000) and microcantilever-linked immuno assays (Moulin et al., 2000; Thundat et al., 2000) to subcellular and cellular micro-electronic systems (Simpson et al., 1998). Analytes measurable by biosensors include a vast array of organic chemicals, biochemicals, inorganics, and metals and more recently ionizing radiation (Lee et al., 2000). Research to integrate microelectronics and nanotechnology (Cunningham, 1998) with elements of gene array technology (Mecklenburg et al., 2000) and cellular engineering may lead to new sensor technology. Bionanosensors are recognized as a key research area in the National Science Foundation's first call for proposals related to the National Nanotechnology Initiative (see http://www.nano.gov/press.htm for details). Such technology could create new capacity for continuous and remote monitoring in chemically and physically complex environmental and structural systems characteristic of DOE's site D&D needs.

Decontamination

Like characterization, decontamination of equipment and facilities is necessary at several stages of the D&D process. Initially, radiation and contamination levels may have to be reduced to allow worker access or to limit their exposure to radiation and other hazards. Decontamination may be required before dismantling or demolition work to prevent the spread of radioactive or toxic materials. Unplanned releases can have off-site as well as on-site consequences. Decontamination procedures are intended to result in a small volume of the most hazardous waste, and much larger volumes of waste that has low or no hazard, thus reducing the cost and long-term risk of disposal. Some decontaminated equipment or facilities might be recycled or reused. The end state of any decontamination activity must be consistent with both site-specific and overall DOE cleanup objectives.

(3) The committee recommends basic research toward fundamental understanding of the interactions of important contaminants with the primary materials of interest in D&D projects, including concrete, stainless steel, paints, and "strippable" coatings.

Rationale: Scientific understanding of the interactions among contaminants and construction materials is fundamental to developing more effective D&D technologies. Both radioactive and toxic contaminants can exist in a variety of chemical forms (for example, in different valence states, complexes, or as colloids), which exhibit very different behaviors. While a good deal of chemical data on the contaminants themselves exist (Delany and Lundeen, 1990), as well as data on their transport in the environment (van der Lee et al., 1997), there is little information of direct relevance to D&D problems. Such information includes how contaminants bind to steel and concrete surfaces, how they penetrate into these materials, their migration into pores, fissures, and welds, and time-dependent "aging" effects (Dzombak and Morel, 1990). Once sufficient thermodynamic and kinetic data on these interactions are obtained to allow their modeling from first principles, the models would allow various decontamination approaches to be evaluated and provide a better way to interpret data from characterization.

(4) The committee recommends basic research on biotechnological means to remove or remediate contaminants of interest from surfaces and within porous materials.

Rationale: The capacity of microbiological processes to destroy, transform, mobilize, and sequester toxins, pollutants, and contaminants is well-established (Young and Cerniglia, 1995). Through research to extend well-known technology in mineral ore leaching and metal recovery (Torma, 1988; Ehrlich and Brierley, 1990), these biochemical capacities may be exploitable for removal of metals and radionuclides from concrete and building debris. An excellent example of which was recently described in an American Society for Microbiology report (see ASM News. 66:133). In addition, microbial biocorrosion processes for structural metals and concrete are well established and the opportunity exists to investigate fundamental processes that could facilitate volumetric reduction of waste from D&D activities. Biotechnical advances in surface treatments of contaminated structures and materials are anticipated from continuing R&D activities, elucidation of biocatalytic properties of biological systems and engineering chemicals, and biosurfactants with unique physical chemical properties (Sullivan, 1998; Banat et al., 2000). A fundamental understanding of the biological processes would also help to ensure that waste by-products from the decontamination could be safely treated and stabilized.

Remote Systems

For D&D work, remote systems provide a unique means to separate workers from hazardous work areas, thus enhancing their safety and productivity. This technology crosscuts all of the other D&D areas—characterization, decontamination, and dismantlement—and has the potential for substantial performance enhancement and cost reduction. The committee recognizes

Dr. Carolyn Huntoon
December 5, 2000
Page 6

the broad range of potential applicability of fundamental advances in this area (Benhabib and Dai, 1991; Andary and Spidaliere, 1993; Gombert et al., 1994; Chen and Burdick, 1995; Paredis and Khosla, 1996; Kapoor and Tesar, 1998; and Tesar, 2000) and makes the following recommendation:

(5) The committee recommends basic research toward creating intelligent remote systems that can adapt to a variety of tasks and be readily assembled from standardized modules. Today's remote systems are one-of-a-kind devices of high cost and limited capability. Their inflexibility leads to rapid obsolescence and is a barrier to their deployment. The recommended initial research focus would be as follows:

    a. Actuators

Rationale: The actuator is the power (muscle) of remote systems, and as such it is the key to performance, reliability, and cost. Except for better construction materials and improved control electronics, most actuator technology has not changed for several decades. Today's actuators typically use only one sensor (for position) so that virtually no real time data (for example, force and velocity) are available to make them "intelligent." More complete sensory input, coupled with decision making software (see below), can produce intelligent actuators that are able to adapt to a variety of tasks. Achieving a relatively inexpensive modular design to allow "plug and play" deployment of these devices would be especially useful in DOE D&D projects because equipment that fails or becomes contaminated is usually discarded. Research to answer the question of granularity (What is the minimum number of required standard modules?) to enable the assembly on demand of the maximum number of remote systems would make the overall system substantially more cost effective in deployment and maintenance.

    b. Universal Operational Software to Provide Criteria-Based Decision Making

Rationale: Criteria-based decision making is the essence of intelligence in robotic systems. It is identical in concept (Miller and Lennox, 1991; Sturzenbecker, 1991; Stewart et al., 1992) to the operating systems in personal computers: What is the best use of the system's resources to perform the task at hand? Today's control of robotic devices is derived from techniques developed during World War II in which control is linear (based only on the difference between two measured parameters). A robot capable of mimicking human adaptability, however, would require a non-linear control system coupled to many parameters corresponding to the physical features that accurately represent performance of the task. The criteria-based software could be universal in the same sense that operating systems on personal computers are universal—one system supports many different applications.

    c. Virtual Presence of the Worker in Hazardous Environments

Rationale: In the initial planning and characterization phases of D&D work, workers often must enter an area of high radiation and contamination that is also congested with left-in-place equipment and materials for which removal inevitably involves physical stress (fatigue) and the potential for personal injury (Fournier et al., 1998). Virtual reality systems could allow workers to perform essential survey and decision making functions from a remote location (Lloyd et al., 1999), thus enhancing their safety and productivity. Advances in the state of the art

Dr. Carolyn Huntoon
December 5, 2000
Page 7

as now used in deep sea exploration should be pursued to improve overall system performance by providing force feedback, remote vision, collision avoidance, and radiation resistant sensor technology.

GENERAL ADVICE

The committee appreciates the quality and openness of presentations it received during its fact finding. This information led the committee to recognize several areas in which it believes advice at this time will be useful to EMSP in its overall strategic planning. The advice summarized below is preliminary, and the committee intends to develop it further in its final report. We note that similar advice has been provided in previous NRC reports and in the Strategic Laboratory Council's recent Adequacy Analysis of the Environmental Quality R&D Portfolio (USDOE, 2000).

(1) The EMSP was initiated by the 104th Congress "...to stimulate the required basic research, development and demonstration efforts to seek new and innovative cleanup methods to replace current conventional approaches which are often costly and ineffective."[10] The committee agrees with the recommendation of earlier NRC reports that EMSP should emphasize basic research (NRC 1997; NRC 1999b). The committee's impression is that most of the EMSP's current efforts are aimed at short term, site-specific D&D problems. Developing new technologies to address current site needs is the responsibility of the DDFA, not the EMSP. While relevance[11] to site-specific problems is important, too narrow a focus may preclude funding valuable "outside the box" research ideas. The NRC Committee on Decision Making made a similar recommendation for some exploratory research to meet the need for backups and alternatives to baseline flowsheets (NRC, 1999c).

(2) To help guide future R&D efforts, the committee believes that DOE-OST should develop a more comprehensive, coordinated, and specific definition of complex-wide D&D needs that could provide a basis for assessing areas in which basic research could contribute. This should include all facilities that EM will eventually D&D under its cleanup mission, whether or not they have been declared as D&D projects. It should identify the major, complex-wide challenges that could give the greatest return on investments in new technologies.

In its final report the committee will provide its own view of DOE's significant D&D problems that cannot be addressed effectively with current technologies, according to its statement of task. Its views will be at an overarching level, based on its one-day fact finding visits to Hanford, Oak Ridge, and Rocky Flats and its review of existing needs lists. The general advice given here and in the next item is intended to help DOE develop a sustainable process for identifying in detail its D&D research needs.

Much of the post-2006 cost savings from new technologies projected by the DDFA is for facilities that have not yet been released to EM (Hart, 2000). The committee saw a variety of lists

---

[10] The program was created in the conference report that accompanied the Energy and Water Development Appropriations Bill (Public Law 104-46, 1995).

[11] Research proposals submitted to the EMSP undergo a two-part review, first for scientific merit and second for relevance to needs identified by Site Technology Coordinating Groups. See point (3) of this discussion for additional information.

Dr. Carolyn Huntoon
December 5, 2000
Page 8

of needs, but they were too narrow, short term, or general to help focus on where basic research could be useful. Developing a complete, useful definition of the D&D challenges will require significant effort from the many DOE facility "owners," but it is essential to developing an effective research and development (R&D) plan.[12] The efforts should start now to help prepare for proposal calls for FY 02 and thereafter.

(3) OST has made an effort to obtain meaningful input from the sites on R&D needs through establishment of the Site Technology Coordinating Groups (STCGs). The committee believes that the DOE contractors and STCGs could contribute more toward identifying true R&D opportunities if they were encouraged by their management to identify major long-term challenges rather than lists of specific technology development needs.

Moreover, there may be value in using an additional, different approach in which members of the scientific community are engaged on a broader sustained basis in understanding the D&D challenges and defining the underlying scientific information that is needed to develop solutions. The new technology or basic research which would effectively address these challenges may be identified more readily by the external scientific/technical community and national laboratories, than by the site operating contractors. Previous NRC reports have recommended greater outreach to universities, national laboratories, and the foreign sector (NRC 1998; NRC, 1999b; NRC, 1999c). The committee will make suggestions for implementing this additional approach in its final report.

(4) The committee believes that EMSP should have a strong role in developing a sound scientific basis for setting standards for the end states[13] of D&D and for release of buildings and materials. The committee also believes this is an appropriate subject for EMSP's R&D program. DOE has no general risk assessment basis for D&D end states. End states for facility D&D are currently, in most cases, being determined on an individual basis, with input on end use from local stakeholders.[14] It is not suggested that this decision process change, but rather that science-based standards be available to the decision makers.

End states for free release of recycle materials are a matter of continuing controversy. The U.S. Nuclear Regulatory Commission has recently asked the NRC for advice on this complex issue.[15] For D&D projects, the end state determines the cost, schedule, and the technology to be used—or even if an adequate technology exists. The importance of establishing end states was noted by the previous NRC committee on D&D technologies (NRC, 1998), and more broadly by other NRC committees[16] and international groups (NEA, 1999; and MacLachlan, 2000).

---

[12] Similarly, a study of DOE's project management practices found that guidelines for planning in the early conceptual and pre-conceptual phases of its projects—when the potential for cost savings is highest—are lacking (NRC, 1999d).

[13] In this discussion, the term "end states" refers primarily to the amounts of chemical or radioactive contamination remaining after a D&D project.

[14] The committee appreciates receiving presentations and comments from Citizens' Advisory Boards during its Hanford, Oak Ridge, and Rocky Flats visits.

[15] Committee on Alternatives for Controlling Release of Solid Materials from NuRC-licensed Facilities, NRC Board on Energy and Environmental Systems.

[16] For example, "An End State Methodology for Identifying Technology Needs for Environmental Management, with an Example from the Hanford Site Tanks" (NRC, 1999b).

Dr. Carolyn Huntoon
December 5, 2000
Page 9

(5) The committee believes that EMSP should consider further encouraging interdisciplinary collaborations, among disciplines such as biology, physics, materials science, and engineering in its calls for proposals.[17] Due to the nature of D&D work, interdisciplinary collaborations are more likely to lead to new, deployable technologies than research in a single area. One early step could be to provide for discussion amongst the proposers of the potential projects selected from review of the FY 01 pre-proposals, perhaps in the context of an EMSP National Workshop such as that held in April 2000.

The committee appreciates this opportunity to make input to the EMSP by this interim report. The final report, which the committee expects to complete in the spring of 2001, will provide more detail and discussion on the recommendations and preliminary general advice presented here.

Sincerely,

Philip Clark, Sr.
Chair

Attachment A: Committee Roster
Attachment B: Statement of Task
Attachment C: List of Reviewers
Attachment D: References

---

[17] The committee understands that EMSP may be able to invest only about $5 million in new D&D research in FY 01, which will restrict the number and size of projects that it can support. The committee's views on EMSP needs for greater funding will be developed in the final report.

Dr. Carolyn Huntoon
December 5, 2000
Page 10

ATTACHMENT A
COMMITTEE ROSTER

COMMITTEE ON LONG-TERM RESEARCH NEEDS FOR DEACTIVATION AND DECOMMISSIONING AT DEPARTMENT OF ENERGY SITES

PHILIP CLARK, SR., CHAIR, GPU Nuclear Corporation (retired), Boonton, New Jersey
ANTHONY CAMPILLO, Naval Research Laboratory, Washington, District of Columbia
FRANK CRIMI, Lockheed Martin Advanced Environmental Systems Company (retired), Saratoga, California
KEN CZERWINSKI, Massachusetts Institute of Technology, Cambridge
RACHEL DETWILER, Construction Technology Laboratories, Inc., Skokie, Illinois
HARRY HARMON, Battelle, Pacific Northwest National Laboratory, Aiken, South Carolina
VINCENT MASSAUT, CEN.SCK, Mol, Belgium
ALAN PENSE, Lehigh University, Bethlehem, Pennsylvania
GARY SAYLER, The University of Tennessee, Knoxville
DELBERT TESAR, The University of Texas, Austin

Staff

JOHN WILEY, Study Director
SUSAN MOCKLER, Research Associate
LATRICIA BAILEY, Senior Project Assistant

Dr. Carolyn Huntoon
December 5, 2000
Page 11

ATTACHMENT B
STATEMENT OF TASK

COMMITTEE ON LONG-TERM RESEARCH NEEDS FOR DEACTIVATION AND DECOMMISSIONING AT DEPARTMENT OF ENERGY SITES

The objective of this study is to provide recommendations to DOE's EM Science Program on the development of a long-term basic research agenda that may lead to new technologies for the deactivation and decommissioning (D&D) of complex, highly contaminated facilities formerly used for the production of nuclear materials. The report will accomplish the following:

- Identify significant D&D problems that cannot be addressed effectively with current technologies.
- Recommend areas of research where the EM Science Program can make significant contributions to solving these problems and adding to scientific knowledge generally.

In recommending specific areas of research, the committee should take into account, where possible, the agendas of other D&D-related research programs.

The committee may also consider and make recommendations, as appropriate, on the processes by which (1) future research needs can be identified, and (2) successful research results can be applied to DOE's D&D problems.

Dr. Carolyn Huntoon
December 5, 2000
Page 12

ATTACHMENT C
LIST OF REVIEWERS

The letter report has been reviewed in draft form by individuals chosen for their diverse perspectives and technical expertise, in accordance with procedures approved by the NRC's Report Review Committee. The purpose of this independent review is to provide candid and critical comments that will assist the institution in making the published report as sound as possible and to ensure that the report meets institutional standards for objectivity, evidence, and responsiveness to the study charge. The review comments and draft manuscript remain confidential to protect the integrity of the deliberative process. We wish to thank the following individuals for their participation in the review of this report:

John F. Ahearne, Sigma Xi and Duke University, Research Triangle Park, North Carolina
Gregory R. Choppin, Florida State University, Tallahassee
Alexander MacLachlan, E.I. du Pont de Nemours & Company (retired), Wilmington, Delaware
Michael Corradini, University of Wisconsin, Madison, Wisconsin
Howie Choset, Carnegie Mellon University, Pittsburgh, Pennsylvania
Mary DeFlaun, Envirogen, Inc., Lawrenceville, New Jersey

Although the reviewers listed above have provided many constructive comments and suggestions, they were not asked to endorse the conclusions or recommendations nor did they see the final draft of the report before its release. The review of this report was overseen by George Hornberger, University of Virginia, Charlottesville, appointed by the Commission on Geosciences, Environment, and Resources, who was responsible for making certain that an independent examination of this report was carried out in accordance with institutional procedures and that all review comments were carefully considered. Responsibility for the final content of this report rests entirely with the authoring committee and the institution.

ATTACHMENT D
REFERENCES

Andary, J.F., and P.D. Spidaliere. 1993. The Development Test Flight of the Flight Telerobotic Servicer: Design Description and Lessons Learned. *IEEE Transactions on Robotics and Automation.* 9(5):664-674. October.

Anderson, G.P., K. D. King, K.L. Gaffney, and L. H. Johnson. 2000. Biosensors and Bioelectronics. 14:771-777.

Banat, I.M., R.S. Makkar, and S.S. Cameotra. 2000. Potential commercial application of microbial surfactants. *Applied Microbiological Biotechnology.* 53:495-508.

Benhabib, B. and M.Q. Dai. 1991. Mechanical Design of a Modular Robot for Industrial Applications. *Journal of Manufacturing Systems.* 10(4):297-306.

Borgermans, P., B. Brichard, F. Berghmans, F. Vos, M. Decréton, K.M. Golant, A.L. Tomashuk, and I.V. Nikolin. 2000. On-line gamma dosimetry with Phosphorous and Germanium co-Doped Optical Fibres. Proceedings of RADECS 1999 conference. Moscow, Russia: IEEE. 0-7803-5726-4/00.

Chen, I-Ming and J.W. Burdick. 1995. Determining Task Optimal Modular Robot Assembly Configurations. *Proceedings of the 1995 IEEE International Conference on Robotics and Automation.* Pp. 132-137.

Cunningham, A.J. 1998. Introduction to bioanalytical sensors. New York, NY: John Wiley and Sons, Inc. 417.

Delany, J.M., and S.R. Lundeen: 1990. The LLNL Thermodynamic Database. Technical Report UCRL-21658, Lawrence Livermore National Laboratory.

Dowling, N.J., M.W. Mittleman, and J.C. Danko, eds. 1990. Microbially Influenced Corrosion. Proceedings of the Symposium-MIC Consortium. Knoxville, TN.

Dzombak, D.A., and F.M.M. Morel. 1990. Surface Complexation Modeling, Hydrous Ferric Oxide. New York: John Wiley and Sons.

Ehrlich, H.L., and C.L. Brierley, eds. 1990. Microbial mineral recovery. New York, NY: McGraw Hill, Inc. 654.

Fournier, R., P. Gravez, and A. Micaelli. 1998. Computer Aided Teleoperation Recent Trends: The CEA Approaches in Mechanics, Control, and Supervision for Maintenance and Dismantling of Nuclear Plants. IEEE International Conference on Robotics and Automation. Louvain, Belgium.

Gilbertson, M. 2000. Telephone discussion with Mark Gilbertson of DOE and the D&D committee, August 24.

Gombert, B., G. Hirzinger, G. Plank, M. Schedl. 1994. Modular Concepts for a New Generation of Light Weight Robots. *Proceedings of the 20th International Conference on Industrial Electronics, Control and Instrumentation. Part 3 (of 3).* Pp. 1507-1514.

Hart, P. 2000. Overview of the Deactivation and Decommissioning Focus Area. Paper presented to the D&D committee in Washington, DC, on March 17, 2000.

Henschel, H., O. Köhn, and H.U. Schmidt. 1992. Optical fibres as radiation dosimeters. Euskirchen, Germany: Fraunhofer-Institut für Naturwissenschaftlich-Technische Trendanalysen, Postfach 1491, Appelsgarten2, D-53864.

Holtz, J.H., and S.A. Asher, 1997. Polymerized colloidal crystalline hydrogel films as intelligent chemical sensing materials. Nature. 389, 829-832.

Huston, A.L., B.L. Justus, and T.L. Johnson. 1996. Fiber-optic-coupled, leaser heated thermoluminescence dosimeter for remote radiation sensing. Appl. Phys. Lett., 68, 3377.

Janata, J. 1989. Principles of Chemical Sensors. Modern Analytical Chemistry. New York, NY: Plenum.

Kapoor, C., and D. Tesar. 1998. A Reusable Operational Software Architecture for Advanced Robotics. Proceedings of the Twelfth CSIM-IFTOMM Symposium on Theory and Practice of Robots and Manipulators, Paris, France, July.

Kress-Rogers, E., ed. 1997. Handbook of Biosensors and Electronic Noses: Medicine, Food and the Environment. Boca Raton, Florida: CRC Press. 695.

Lee, C.W., J. Min, S.H. Moon, R.A. LaRossa, and M.B. Gu. 2000. Detection of radiation effects using recombinant bioluminescent *Escherichia coli* strains. *Radiation Environment*. Biophysics. 39:41-45.

Lieberman, R.A., ed. 1996. Chemical, Biochemical and Environmental Fiber Sensors VIII. Proceedings of SPIE (Society of Photo-optical Instrumentation Engineers). Vol. 2836.

Lloyd, J. E., J.S. Beis, D.D. Pai, and D.G. Lowe. 1999. Programming Contact Tasks Unisn a Reality-based Environment Integrated with Vision. IEEE Transactions on Robotics and Automation. June1999. 15(3):423-434.

MacLachlan, A. 2000. Radiation experts don't agree on how to release materials. Nucleonics weeks, 41:13, March 30.

Mecklenburg, M., B. Danielsson, H. Boije, I. Surugui, and B. Rees. 2000. "DNA-Based Biosenors: A Tool for Environmental Analysis." In: A. Mulchandani and O.A. Sadik, eds. Chemical and Biological Sensors for Environmental Monitoring. American Chemical Society, Washington, D.C. Symposium Series 762. p310.

Miller, D. J., and R.C. Lennox. 1991. An Object-Oriented Environment for Robot System Architectures. IEEE Control Systems. Pp. 14-23. February.

Miller, R.W. et al. 2000. Dosimetry characteristics of an optically stimulated luminescent fiberoptic dosimetry system. Paper presented at the 2000 World Congress on Medical Physics and Biomedical Engineering, Chicago, IL, 23-28 July.

Moulin, A.M., S.J. O'shea, and M.E. Welland. 2000. Microcantilever-based biosensors. Ultramicroscopy. 82:23-31.

NEA/IAEA/EC. 1999. Proceedings of the NEA/AEA/EC Workshop on the regulatory aspects on decommissioning. Rome, Italy. May.

NRC. 1996. Affordable Cleanup? Opportunities for Cost Reduction in the Decontamination and Decommissioning of the Nation's Uranium Enrichment Facilities. Washington, DC: National Academy Press.

NRC. 1997. Building an Effective Environmental Management Science Program: Final Assessment. Washington, DC: National Academy Press.

NRC. 1998. A Review of Decontamination and Decommissioning Technology Development Programs at the DOE. Washington, DC: National Academy Press.

NRC. 1999a. An End State Methodology for Identifying Technology Needs for Environmental Management, with an Example from the Hanford Site Tanks," Washington, DC: National Academy Press.

NRC. 1999b. Research Needs in Subsurface Science. Washington, DC: National Academy Press.

Dr. Carolyn Huntoon
December 5, 2000
Page 15

NRC. 1999c. Technologies for Environmental Management. Washington, DC: National Academy Press.

NRC. 1999d. Improving Project Managment in the Department of Energy. Washington, DC: National Academy Press.

Paredis, C. and P. Khosla. 1996. Kinematic Design of Serial Link Manipulators from task Specifications. *International Journal of Robotics Research,* 12(3):274-287.

Simpson, M.L., G.S. Sayler, B.M. Applegate, S. Ripp, D.E. Nivens, M.J. Paulus, and G.E. Jellison Jr. 1998. Bioluminescent-Bioreporter Integrated Circuits form Novel Whole-Cell Biosensors. *Trends in Biotechnology.* 16:332-338.

Stewart, D.B., R.A. Volpe, and P.K. Khosla. 1992. Integration of Real-Time Software Modules for Reconfigurable Sensor-Based Control Systems. *Proceedings of IEEE/RSJ International Conference on Intelligent Robots and Systems.* Pp.325-332. July

Sturzenbecker, M.C. 1991. Building an Object-Oriented Environment for Distributed Manufacturing Software. *Proceedings of IEEE International Conference on Robotics and Automation.* Pp. 1972-1978. April.

Sullivan, E.R. 1998. Molecular genetics of biosurfactant production. *Curr. Opin. Biotechnology.* 62:1-34.

Tesar, D. 2000. A National Program to Deploy Robot Systems for Hazardous Waste Site Clean Up. International Decommissioning Symposium sponsored by DOE and IAEA, June 12-16, Knoxville, Tennessee.

Thierry, D., ed. 1997. Aspects of Microbial Induced Corrosion: Papers from Eurocorr '96 and the Efc Working Party on Microbial Corrosion. European Federation of Corrosion. 162.

Thundat, T.G., K.B. Jacobsen, M.J. Doktycz, S.J. Kennel, and R.J. Wamack. 2000. Micromechanical antibody sensor. PCT Int. Appl. CODEN:PIXXD2 WO0058729.

Torma, A.E. 1988. Leaching of metals. In: H.J. Rehm and G. Reed, eds. Biotechnology. Weinheim, GR: VCH Verlagsgesellschaft. 6b:367-399.

Updike, S.J. and G.P. Hicks. 1967. The Enzyme Electrode. *Nature,* 214:986.

USDOE. 1998a. Accelerating Cleanup: Paths to Closure. DOE/EM-0362. Washington, DC: USDOE.

USDOE. 1998b. GammaCam Radiation Imaging System, Innovative Technology Summary Report. OST Reference #1840. Washington, DC: U.S. DOE. February.

USDOE. 1999. Long Range Alpha Detection for Component Monitoring, Innovative Technology Summary Report. DOE/EM-0497. Washington, DC: USDOE. September.

USDOE. 2000. Adequacy Analysis of the Environmental Quality Research and Development Portfolio. Draft revision b, July. Washington, DC: USDOE.

van der Lee, J., E. Ledoux, G. de Marsily, A. Vinsot, van de Weerd, Leijnse, A., Harmand, B., Rodier, E., Sardin, M., Dodds, J., and Hérnandez Benitez, A. 1997. Development of a Model for Radionuclide Transport by Colloids in the Geosphere. European Commission Nuclear Science and Technology. EUR 17480.

Van Hecke, G.R., and K.K. Karukstis. 1998. A Guide to Lasers in Chemistry. Jones and Bartlett Publishers.

Webster, J.G., ed. 1999. The Measurement, Instrumentation and Sensors Handbook. Boca Raton, FL: CRC Press LLC.

Young, L.L. and C.E. Cerniglia, eds. 1995. Microbial transformation and degradation of toxic organic chemicals. New York, NY: John Wiley and Sons, Inc. 654.

D

Illustrative Science Base and Scope for Remote Technology for Decontamination and Decommissioning of DOE Nuclear Facilities

Background

The decommissioning and dismantlement of nuclear facilities entails a wide range of physical tasks (e.g., inspection, cutting, handling, packaging) of a very diverse set of structures (e.g., piping, valves, pumps, tanks, wire conduits, building structures, concrete). Successful D&D activity is possible today using existing technology based primarily on manual operations, which entails high cost and long timelines. Numerous attempts have been made to use remote systems technology (robotics) to reduce hazards to workers and to reduce the cost of the total life cycle of the operation. Barriers to the deployment of an advanced technology arise from a history of taking a low-risk approach, the deployment of industrial robotics technology poorly suited to the functional spectrum associated with D&D, or the development of on-off devices (with a nominal record of performance).

All of this disparate activity suggests the absence of a national strategy (based on adequate resources) for a science base that would aggressively attack the technology needs with a balanced approach where timely investment can be made to address weaknesses and opportunities to most rapidly deploy a cost-effective technology. The existing project-oriented approach results in special-purpose systems developed under contract primarily by industry to meet a local site need independent of long-term research activity at universities and the DOE's own research laboratories. Further, this special-purpose technology depends on a special class of maintenance technicians provided by the system supplier therefore severely limiting operational flexibility at any given site. By contrast, with a national strategy, generalized technology could be deployed to not only cover all relevant applications but could do so at low risk and dramatically reduced costs. In fact, it is suggested that deployment costs can be reduced by more than 50 percent, and using sophisticated training and operational software, timelines could be substantially reduced, and overall logistics functions (maintenance) also could be dealt with as a cost-cutting opportunity.

How Can This Be Done?

First, central computer controllers now have an open architecture and their cost is coming down rapidly while their performance is increasing. The expected controller cost with software for the open architecture remote systems is expected to be based on continuously advancing personal computers and to cost less than $10,000.

Second, even though a very broad range of remote systems will be required (handling, manipulators, rovers, etc., all at various scales), they can be built with an open architecture based on a collection of about ten standard actuator modules with an average cost less than $5000 in reasonable quantities (100+). These actuators are 80 percent of the remote system. They contain all the wiring, motors, brakes, local electronics, sensors, interfaces, etc. Because of their standardized interfaces, they can be rapidly assembled on demand to create a wide population of remote systems to meet any specific application.

Third, because of the standard interfaces, the actuators can be removed and rapidly replaced and a minimal set of spares kept on hand to maintain the system. Also, expensive, highly trained service personnel become a thing of the past (just as has occurred for computer systems). Hence, life-cycle costs come down a great deal. These standardized interfaces would also apply to sensor modules (used for site characterization), process tools (for size reduction), and to communication subsystems (power umbilicals, wireless data transfer, etc.).

Fourth, because the remote system is made up of modules having standardized interfaces, tech mods (better motors, sensors, brakes, embedded electronics, controllers, etc.) can be inserted to improve the subsystem without disturbing the standards and done so at very low cost. Hence, the up-front risk to the designer, project decision maker, and site contractor goes way down. Once the standards are accepted, in-depth qualification tests of these subsystem modules becomes possible, certifying their performance metrics for a wide range of operational conditions. This is equivalent to the lessons learned from the market for personal computers: performance goes up while costs go down.

Fifth, it is now possible to establish a universal operating software to operate all remote systems that can be built from the standardized subsystems. This is a lesson learned from Microsoft. We need to apply it to mechanical systems as well as for computer-based systems. The robotics research community has in place a broad body of analytics to support decision making required to operate complex systems (from 1 to 20 degrees of freedom), to expand their performance envelope, and to allow the judgment of the human operator to efficiently manage this complex system through a spectrum of organic interfaces.

Economics and Program Planning. D&D represents an estimated life-cycle cost to DOE of $33 billion, of which only 12 percent will be performed before FY 2007. Hence, there is a long timeline in which to properly develop cost-effective technology. A good long-term program for development of remote systems has been developed by the DOE community and others in the form of the Robotics and Intelligent Machines (RIM) roadmap. This program lays out a strategic timeline for overall performance goals and targets for a demanding list of D&D applications (as well as mixed waste, nuclear materials, etc.). One of these performance goals is to reduce worker exposure to radiological hazards by 90 percent while increasing productivity by 300 percent for such applications as the Hanford building 324 hot cells, the Rocky Flats glove box size reduction, etc.

This plan recognizes the need for advances in certain component technologies (planning in unstructured environments; characterization sensor suites, data analysis, management, and visualization; specialized manipulators and handling equipment; mobile platforms and their navigation; configuration management; man-machine interface and operator training; tetherless operation; etc.). However, the RIM plan does not offer a science strategy to respond to these valid program goals.

Remote Systems Science Activity To Support Long-term Development for D&D.

The following list of topics provides a reasonable description of the technologies and the required science to support advanced remote system development for D&D. Each is described in terms of the most promising development and, in many cases, represents a major transition from present research in the topic.

Electronics Hardening. Semiconductors continue to undergo intense development of smaller geometries, higher doping, reduced operating voltages, ultra-thin active regions, reduced dielectric strengths, and developments in device structures (hetero-junction bipolar transistors, high-electron mobility transistors, quantum well, super lattice, v-groove, etc.), which offer many improvements for radiation tolerance. In-depth understanding of the device physics must be developed to create models as the basis for the design of new technologies for radiation tolerance. Empirically, we can confirm that newer electronic components are more tolerant to terrestrial radiation (total dose) environments but at the expense of a decreased Single Event Phenomena (SEP) capability (important in low-radiation fields).

As geometries become smaller, multiple errors as a result of a single ion strike and new potentially destructive mechanisms such as a rupture of the oxide layer may cause operational problems. A challenge exists to develop a science base that would predict when a device will no longer tolerate the multiple effects from radiation, especially as the complexity of these devices increases. Recent development of MESFET GaAs technology and bipolar technologies suggest that total dose radiation tolerance will be greatly improved in the future. For the system designer, it becomes essential to have explicit knowledge of the radiation tolerance of any given device, now realizable only by continuous testing and evaluation. Hence, a science of tolerance measurement of a broad range of devices, the acquisition of lessons learned, and the reduction of these data as guidelines to the electronics designer now becomes highly desirable.

Sensors for Characterization and for System Operation. Here, the principal objective is to create a new level of science to accelerate the technology deployment for the mobile and remote mapping of large indoor and outdoor areas consisting of pipe and tank-like structures. This includes the processing of sensor data to identify objects within the acquired scenes (including radiation levels and sources) as well as aid in the visualization of the scene by the operator. These technologies include data registration, multi-sensor data fusion, scene building, object and scene modeling, object detection and recognition, and sensor characterization (see Chapter 4). Specific research tasks might be listed as follows:

Task	D&D Application
1. Next Best View in 3D Object Modeling	Uses only a few views, without prior knowledge, permits object modeling. Use as aid to robot in identifying unknown objects in D&D structure to be dismantled.
2. Registration of Multi-Sensory Modalities	Automatically register multi-modal 2D projective data sets of a scene to a global/3D coordinate system (including range and color data). Useful for mapping robot's work environment and providing collision data for remote operation of multi-degree of freedom mobile robot and manipulator.
3. Multiple View, Multiple Wavelength, Thermal and Radiation Imaging for 3D Object Characterization	Simulated data capture for thermal and radiation sources, construct time varying scenes as the source levels vary. Data can be mapped as a texture onto 3D object scenes. Multiple wavelength infrared and radiation cameras permit true temperature and radiation data capture without tedious calibration.
4. Multi-modal Measurement for Image Collage	Register multiple range and multi-modal images to a common reference frame, integrate into a unified, textured, 3D model for display. Permits D&D scene construction to guide the robot for inspection and sequential structure dismantlement.
5. Data Reduction of 3D Meshes for Multi-Resolution Analysis Using Wavelet Transform	Use results of multi-resolution analysis to guide a mesh reduction strategy, use quincunx wavelet transforms. D&D scenes can produce data magnitudes that can be overwhelming. This approach retains only sufficient data to enable practical visualization and processing.
6. Object Modeling in Multiple Object Scenes Using Deformable Simplex Meshes	Automatically models 3D point clouds that might comprise multiple objects using deformable simplex meshes, shrinking the mesh according to physical constraints, then refining the mesh to identify multiple objects allows determining the position and orientation of complex pipe structures, valves, even fine segmentation of small objects as can be expected in the D&D work environment.
7. Volumetric Primitive Object Presentation of Range Images for Object Recognition	Recover geometric models for each part of each object from range image data, reduce noise and background clutter, identifying objects with finite number of primitive models. Enhance robot operations on repetitive (similar objects) removals from the D&D environment.

A p p e n d i x D

Mobile Platform Navigation and Operation. Position estimation deals with the ability of a mobile robot to estimate its position relative to the environment (both known and unknown). GPS provides typical accuracy of 10 meters which has little value for D&D operations and is altogether not usable indoors. For this reason, scientific development is required on systems employing external references (such as beacons and artificial landmarks) and on dead-reckoning systems. To date, all existing navigation systems have distinct shortcomings for use on D&D.

Obstacle avoidance is a serious problem. The need is to detect obstacles in time to avoid collision and to circumnavigate the obstacle. Research is warranted in two main areas: computer vision-based systems (i.e., those that employ cameras) and those that employ range sensors. The focus in computer vision performance is suggested to be on advanced computer software. For range sensing, continued development of LIDAR and FLASH LIDAR sensors (using range snapshots providing rich information for all objects within sight of the sensor) seems promising.

Dramatic improvements in the kinematic design of mobile robots will be necessary before truly versatile performance can be expected. In unstructured environments (as found in D&D), it is quite difficult to be fully functional on wheels only (wall climbing, internal pipe operation, traversing rubble, etc.). Alternative methods of propulsion are usually less efficient and incur penalties in weight, complexity, control, and accuracy. For example, legged robots are slow, fragile, and cumbersome in operation. Suggested research would include a reconfigurable system that could be rapidly assembled from standard modules, enabling the utilization of several distinct modes of propulsion or even combinations.

Virtual Reality for Enhanced Man-Machine Interface and Training. In the initial planning and characterization phases of D&D work, workers must enter an area of high radiation and contamination that is also congested with left in place equipment and materials for which removal inevitably involves physical stress (fatigue) and the potential of personal injury. Virtual reality systems combined with mobile robot platforms (including advanced navigation technology) could allow workers to perform essential survey and decision making functions from a remote location, thus enhancing their safety and productivity. Advances in the state of the art as now used in deep sea exploration (navigation, scene analysis, specialized sensors) should be pursued to improve overall system performance by means of force feedback (of physical operations), remote vision (for registration, object locations, radiation sources, etc.), collision avoidance (because of the lack of definition of the work area), and radiation-resistant system technology (especially the sensors themselves). The required science involves an enduring issue of a balanced technology to enhance the relationship between the operator and the machine. This is partially due to the increasing complexity (options and performance) represented by the machine and the desire to perform more complex tasks. The required science must deal with the following questions. What are the most efficient channels of communication (visual, sound, kinesthetic), which channels can easily be overloaded (too much information flow not easily interpreted by the operator), which can be used to reduce the potential for operator fatigue (both physically and mentally), enhance training, provide in-situ skilling, and overall can this technology reduce costs specifically for the characterization and physical tasks associated with D&D.

Intelligent Actuators. We all recognize the pervasive influence of the computer chip in modernizing our information technology. The equivalent in intelligent machines is the actuator. We need to embed the same level of excellence in component and system technologies in the actuator as we now do in the component technologies for computers. Only then will we dramatically improve performance while reducing costs. Only then can we consider building fully integrated machines on demand (as we now do for computers). Here, we recommend a concentration in all the sciences to modernize actuator technology so that it stands as an equal partner to the computer chip in order to create a balanced technology for a broad population of remote systems for D&D. The following listing of scientific concentrations is recommended to dramatically impact the level of technology for intelligent actuators.

1. **Advanced Component Technology** —One of the most troublesome components in an actuator is the gear train; it needs to be modernized to be lighter, stiffer, easily manufactured, etc. Similar efforts are required for brakes, bearings, prime movers, magnets, wiring insulation, etc. Also, standardized quick-change interfaces of high precision need to be developed of high rigidity and lowered cost.

2. **Micro-Sensors** —Unfortunately, almost all existing actuators contain only one sensor, making it impossible to make the module intelligent. Here, we recommend the embedding of 10 (or more) sensors in every actuator (temperature, torque, current, voltage, position, etc.) to generate a full awareness of the actual condition of the actuator. These sensors should benefit from

advanced MEMS technology and represent the best production experience now used to make sensors for automobiles (i.e., very small, rugged, low cost, etc.).

3. Actuator Design —Literally hundreds of parameters will be involved in the design of these actuators. A process must be put in place to interactively control these parameters so that advanced structures will result to meet the most demanding application requirements (extreme low weight, high power density, good disturbance rejection, etc.). This includes the design of the prime mover, choice of materials, method of fabrication, etc.

4. Operational Software —Present actuator technology depends primarily on an outdated concept of feed-back control. What is needed are criteria-based decision making systems to maximize performance, to make it possible to provide condition-based maintenance for maximum system availability, and to provide a fault tolerant architecture to reallocate resources inside the actuator to continue system operation even in the presence of a fault.

5. Maximum Performance —Most actuators are conservatively operated to prevent saturation. Here, we wish to maximize performance just as we do for the engines of our fighter aircraft. It is recommended to test each actuator as-built, document all parameters, useful operational criteria, their norms, combinations of criteria to meet specific performance objectives, conflict resolution among performance objectives, etc. All of these issues should be prioritized by the operator in the field, archived for lessons learned, and to advise the operator relative to the performance reserve available from a given actuator.

6. Layered Control —Virtually all present actuators operate at only one scale. The biological system has been shown to operate at three to four distinct levels of motion. It is recommended to create a similar set of scales in the full architecture of these actuators, making it possible to maintain a much higher level of accuracy even under external disturbances. Special use of MEMS technology will make layered control possible.

7. Ultra-Light —Frequently, applications occur where actuators of considerable torque capacity must be extremely light. It is now possible to aggressively reduce the weight of these actuators, while still making them shock resistant as well as extremely compact. The best selection of materials, optimum selection of prime mover technology, and highly structured design rules and procedures can address these issues to meet this objective.

8. Fault Tolerance —Normally, fault tolerance is achieved by duplication of actuators—one dormant while the other operates—a complete disaster for efficiency, weight, and cost. Here, it is recommended that a dual system fully integrated be developed where all components are used at all times for maximum performance with no single point failures. Should a partial failure occur, the system would be reconfigured based on criteria to minimize the affect of the failure and still achieve nearly optimum performance. This capacity is essential where access to the remote system is difficult and where endurance and long-term availability are paramount.

9. High Efficiency —Certain applications demand the use of the absolute minimum of power, primarily to minimize the weight and cost of portable power packs. Generally, this requirement implies very little heat generation by using criteria to reduce spurious responses, peak torques, hysteresis in magnetic materials, etc. An intelligent actuator based on 10 sensors is the ideal technological environment in which to treat extraordinary requirements of this class.

Universal Operating Software for D&D Systems. The suggestion here is to develop the science base for universal operational software for mechanical systems that are increasingly becoming more intelligent in response to a need for greater performance, flexible response to changing requirements, fault tolerance for continued operation under a fault, and condition-based maintenance to enable repair by rapid module replacement by a nominally trained technician—all to be achieved at reduced costs.

The crucial reality is that a gigaflop controller technology exists today as a $5000 commodity to provide immense decision making resources in real time to allocate resources within increasingly complex mechanical systems. By the year 2010, it is predicted to be 50 gigaflops. This means that the simple analog concept for "control" of our mechanical systems must now give way to a sophisticated criteria-based "management" of excess resources (options) to maximize the system's response to human intervention and to create open architecture systems that can be rapidly reconfigured to respond to a fault, to integrate a new technology, and to maximize performance (at reduced cost). Elements of this approach have been pursued for our fly-by-wire aircraft, automobiles, household appliances, etc. In order that we create the most flexible, responsive, and cost-effective technology for D&D, it is essential that these unique software requirements be met to augment the science of the software field itself.

Increasingly, our most advanced mechanical systems are being controlled by a predictive model reference based on as-built parameters to maximize their performance. Actual performance is measured by distributed sensors (both internal and external to the structure) to provide information about functions at all levels (i.e., the sensor model). The difference between the modeled and sensor references provides a residual as the basis for criteria-based digital control. This makes the following functions possible:

Enhanced Performance: A functional map of the system makes it possible to predict its performance (in terms of a series of criteria) to meet a very broad range of objectives (rapid response, disturbance rejection, high load, etc.).

Condition-Based Maintenance: The difference between the model reference and the sensor reference can be used to identify failure of parts within the system and to plan for repairs and tech mods.

Fault Tolerance: Excess resources in the system (in every active component such as motors, brakes, gear trains, etc.) allow the system to maintain operation by isolating the fault and reconfiguring the system to achieve the desired level of performance.

The required decision making criteria are unique to the mechanical domain. The requirement for openness in the architecture of the mechanical structure creates the potential for dramatic improvements in cost/performance ratios (as has occurred over the past three decades for computers). A fundamental need is universal operating systems (based on a scientific development of the operational software) in the same manner as demonstrated by Microsoft for personal computers. Because the mechanical system's domain is highly complex and nonlinear, the software must adapt to this complexity. Hence, the science must move towards this opportunity in order to support the development of the general field of intelligent mechanical systems that is just now emerging.

To achieve this level of performance will, indeed, be demanding. Whenever we have invested heavily in any major technology (fighter aircraft, nuclear reactors, submarines), we always invest heavily in the role of the operators and their training in order to maximize the effectiveness of the whole system. This is certainly true for D&D because of the dominant role of the operator. The primary development for the required software is given in the following:

1. **Configuration Management** — This includes complex questions of resource allocation for improved obstacle avoidance, arm stiffness, end-effector dexterity, etc., while prioritizing alternate configurations (tools, power packs, controllers, actuators, etc.) to the decision maker (the D&D system operator).

2. **Obstacle Avoidance** — Develop a mathematical formulation for potential field forces between an array of known obstacles and the structure of the robot. Criteria can be developed to best guide the system through an obstacle-strewn environment (or to find a target) with kinesthetic feedback to the operator through a haptic interface.

3. **Criteria Development** — Criteria are expressed as mathematical descriptions of the physical attributes of the robot (for example: load capacity, energy consumption, accuracy, etc.). Fundamental to the success of this effort is the development of norms for these criteria and their physical meaning to the operator.

4. **Criteria Fusion** — Combinations of these criteria (fusion) become essential in the formulation of task performance indices to enable the decision maker to obtain maximum performance. The operator must be given the opportunity (either through insight or through training) to prioritize (rank) these criteria, to ask for certain combinations to form indices on demand, and to learn what works best for a given task.

5. **Fault Tolerance** — Criteria-based control allows for the selection of the best configuration of the robot to minimize the impact of any given fault (which will usually occur as a reduced load capacity, accuracy, responsiveness, etc., in an actuator) and still maintain operation.

6. **Operational Software** — The required decision making software for management of all available resources is far beyond any standard control approaches (PID, fuzzy logic, adaptive control, etc.). This means that a revolutionary software architecture is required which is not only modular (object oriented) but also extensible, reusable, and operates in real time.

7. **Man-Machine Interface**

The operation of very complex systems (dual arms, multiple slaves, remote control) to perform demanding tasks (disturbance rejection, handling of ungainly objects) is best achieved by setting operational priorities (selection of criteria, performance indices, threshold levels for fault identification, etc.) by human intervention.

Condition-Based Maintenance (CBM). Perhaps the most important development to gain acceptance of remote systems is to assure a high level of availability of that system, to make it rapidly repairable by onsite personnel, and therefore, to substantially reduce forced outage time and the cost of operation. For example, a system that would detect and report that an actuator had degraded to 80% of its performance.

Then, the operator (or D&D manager) would evaluate its impact on the spares available at that time and the potential impact on the dismantlement schedule. To accomplish this objective, it becomes necessary to parametrically model each subsystem (all mass, deformation, friction, electrical resistances, etc.), describe its performance (response, load capacity, linearity, etc.) in terms of a selected set of criteria (the hardest problem), establish the effects of their performance on the overall performance of the larger system, and to embed this decision process partially at the subsystem level (actuators, sensors, communication nodes, etc.) and to download the evaluation results on a disk or directly to the manager's control station.

Selecting and defining the physical meaning of a set of performance criteria, both at the subsystem and the system levels, is the most critical issue and would require intense involvement by the D&D community. If it were done well, the operator would be emancipated from the uncertainty associated with maintenance, false alarms would go way down, sudden failures would greatly disappear, and the cost of the overall operation would go down. The project planner would be able to more accurately predict the availability of all the systems, dramatically reducing the threat of major shutdowns and improving safety. The managers would know at all times what systems were at 100 percent, 80 percent, or even 50 percent of their performance levels. They would know what each subsystem's performance meant to what is important to the operation (safety, project schedule, logistics spares, etc.). They would then decide when to maintain the subsystem, how long it would take, should they wait until a scheduled maintenance period, etc.

Identification of D&D Task Parameters. In order to define and prioritize a long-term science program for remote systems for use in D&D, it is essential to have a clear parametric definition of the physical tasks to be performed. This type of data can only be aggregated by careful analysis and by dedicated personnel. The existing needs documents fall far short of the numerical clarity that is necessary to plan a two-decade-long research activity. Detailed, quantitative information could be obtained by in-depth analysis of a few of DOE's most significant D&D tasks (i.e., PNNL—Building 324; Rocky Flats—glove box size reduction; ORNL—Building 3019; and INEEL—Engineering Test Reactor). It would be well to also include some data on the D&D task requirements from our commercial nuclear reactors. Examples include a standard project description; a parametric description of distinct physical tasks for that project; planning experience involving technology transfer, operator training, and cost effectiveness; deployment issues associated with the proposed technology; and comments on the requirements to drive the needed science. The in-depth analysis of the applications chosen should include distinct information that should be provided such as:

- timeline and frequency of physical tasks
- duration of each task
- parameters describing the task: geometry, forces, speeds, accuracy, required dexterity, etc.
- handling requirements
- record documentation requirements
- radiation levels expected
- level of direct operator involvement

E
Acronyms

AEC	Atomic Energy Commission
ALARA	as low as reasonably achievable
BRWM	Board on Radioactive Waste Management
COLEX	column exchange
D&D	deactivation and decommissioning
DDFA	D&D focus area
DCGL	derived concentration guideline level
DOE	Department of Energy
EC	European Commission
ELEX	electric exchange
EM	Office of Environmental Management (at DOE)
EMSP	Environmental Management Science Program
EPA	Environmental Protection Agency
GFP	green fluorescent protein
HEU	highly enriched uranium
IAEA	International Atomic Energy Agency
ISTC	International Science and Technology Center
LA/ICP/AES	laser ablation inductively coupled plasma atomic emission spectrometry
LA/MS	laser ablation mass spectroscopy
LEU	low-enriched uranium
LIBS	laser-induced breakdown spectroscopy

MEMS	micro-electro-mechanical systems
NABIR	Natural and Accelerated Bioremediation Research
NEA	Nuclear Energy Agency
NRC	National Research Council
OECD	Organization for Economic Cooperation and Development
OSL	optically stimulated luminescence
OST	EM Office of Science and Technology
PCBs	polychlorinated biphenyls
PUREX	plutonium uranium extraction
R&D	research and development
REDOX	reduction oxidation
RIM	robotics and intelligent machines
SRS	Savannah River Site
STCGs	site technology coordinating group
TCE	trichloroethylene
TL	thermal luminescence
TLD	thermoluminescence dosimetry
USNRC	Nuclear Regulatory Commission